Cristian Angeli – **BUILDING SAFETY, EFFICIENCY AND COST SAVINGS**
Scientific studies on ICF building system Insulating Concrete Forms

ISBN – 978-88-31668-54-5

Youcanprint
Via Marco Biagi 6, 73100 Lecce
www.youcanprint.it
info@youcanprint.it

NOTE

The Author assumes no responsibility or liability for any errors in the content. Any action the reader takes upon the information in the present book is strictly at their own risk.
The Author assumes no responsibility or liability for any direct or indirect content that may be found in the website references listed in Chapter 3. Any action the reader takes upon the information in those web pages is strictly at their own risk.
After carrying out the proper research on the rights relating to the illustrations contained in the present book, the Author is nonetheless available for any remuneration due to those entitled.

Cristian Angeli

BUILDING SAFETY, EFFICIENCY AND COST SAVINGS

SCIENTIFIC STUDIES ON
ICF BUILDING SYSTEM
INSULATING CONCRETE FORMS

English version

TABLE OF CONTENTS

Introduction

The first rudimentary attempts at a construction technique resembling what is now known as Insulating Concrete Forms dates back to sixty years ago. However, the diffusion of "modern" ICFs - that is to say insulating formworks of ETS (usually polystyrene) - began in the 1980s in the U.S.. ICFs rose quickly all over the United States until they became one of the most popular construction systems for residential buildings due to their higher efficiency than the conventional wooden frames. As a direct consequence, the United States saw the spread of ICFs and related products manufacturers. Along with the specialized installers, they have played a fundamental role in the American economy ever since.

In Europe, and particularly in Mediterranean basin, ICFs saw a slower rise. Despite a few products already manufactured in Switzerland during the 1970s, unconventional and innovative construction methods among whom ICFs have been opposed for decades. This was mainly due to the workers' skepticism towards new construction technologies since the classical methods were more rooted in the social texture than they were in the U.S. Therefore, wood and brick are to the present day the most widespread construction materials in Northern Europe, whereas Southern Europe still employs mainly conventional reinforced concrete. Nothing has really changed ever since the post-war period...

This delay was due to a number of complex reasons. On one hand, it was strongly determined by our culture since the European concept of "house" as a "home" has roots that run deeper here than in the United States. On the other hand, many designers might have been discouraged by the old-fashioned prescriptive codes existing.

However, this makes nothing more than a simple delay. Over the past decade, the building industry has undergone a technical as well as social paradigm shift in Europe and especially in Italy.

This paradigm shift is attributable to two main factors. The first factor is the introduction of new yet prescriptive technical regulations that substantially raised the performative standards requirements of energy efficiency and seismic safety for buildings. The second factor is the persistent economic downturn Italy is undergoing. Due to the economic crisis, everyone now from the builder to the final user needs to "make it sooner" and "faster" while saving money. The search for new construction solutions has therefore led to the uncovering of the brand-new excellence and consequently to the gradual launch of ICFs in Europe.

As previously stated, the credit goes namely to the "search" for new solutions through research and experimentation - which brings us to the purpose of this work. This book aims to underline the main scientific studies carried out in the field of ICFs all over the world. Oddly enough, these works are hardly known to the professionals operating in the field themselves. If taken individually, none of these pieces of research seems substantial, although all together, they do provide a full overview on this effective technology.

The present study is the result of a systematic web search through specific search engines. Fifty works including books and scientific papers concerning ICF systems have been selected. All studies were carefully read and analyzed by the Author of this book in order to give an easily accessible summary of each one. The target of this book are all those engineers and researchers who would like to get a glimpse of these systems. However, this study does not aim to replace the reading of the original works, but rather to provide documentary evidence on their existence as well as to summarize their content. For this reason, every summary provides the full web references to the original work.

The search for articles was focused on those namely concerning the properly so-called ICF systems - that is to say, the flat wall type composed of small polystyrene formworks. This means that studies on ICFs made of "big forms", "grid walls", and other materials different than polystyrene were not included in this collection.

It is worth pointing out that these researchers, who have worked independently over the four corners of the earth, from India to Romania and from Italy to the U.S., all came up with different and sophisticated solutions. This leads us to the conclusion that such a mature system as ICFs is nonetheless so flexible as to leave plenty of room to creativity and experimentation.

On the contrary, it is quite sad to observe how little European Universities are active on the topic compared with the American ones.

Hopefully enough, this unprecedented mosaic of precious scraps of knowledge will fill this cultural vacuum by making this specific scientific background accessible to all. Furthermore, it might help this responsive to present-day needs construction method take increasing hold.

After all, ICF construction allows to combine energy efficiency with faster work on site, structural safety and cost savings - as indisputably shown in this book.

Cristian Angeli

Acknowledgements

A special thanks goes to Bazzica Group based in Trevi (Italy), a leading company in expanded polystyrene production and manufacturer of the ICF ITALIA construction system (www.icfitalia.eu), for their invaluable contribution to the present work. Many of the pictures of poured in place forms were put at our disposal by them.

Further acknowledgment for the photographic material goes to the ICF ITALIA partner companies:

-   ICF-France based in France (www.icf-system.com)
-   ICF EFGAD based in Israel (www.icf-efgad.co.il)
-   TECdream based in Portugal (www.tecdream.com)

Translated by Sofia Guerra

CHAPTER 1

ICF Construction System

1.1) Overview

ICF systems are composed of EPS forms (sintered Expanded Polystyrene) and used to build reinforced concrete walls with high thermal insulation. Once the concrete is poured in place, this construction system allows the construction of antiseismic, comfortable and highly sound insulated buildings.

There are many and various ICF systems spread around the world, especially in the U.S.. However, the Author of the present book selected one ICF typology that he considers to be the most high-performant, as it combines the greatest design and construction benefits. This specific ICF system is characterized by some properties that help minimize the construction costs while fully complying with the current European technical regulations.

The system referred to is the ICF ITALIA system, produced by the Italian manufacturer Bazzica, based in Trevi (Italy). The Bazzica group formworks will be frequently mentioned over the subsequent descriptions as well as in the illustrated section. Further information on their work is available at www.icfitalia.eu.

Given the proper adjustments, however, the following considerations could apply to the majority of existing ICFs with "small forms" and expanded polystyrene just as fittingly.

The ICF ITALIA formworks have a specific structure which guarantees the highest thermohygrometric comfort during both summer and winter as they provide a high thermal insulation complying with the current regulations. Also, the uniform flat surface of the materials and the respective calibrated thickness prevent the formation of thermal bridges on wall junctions.

The ICF ITALIA system is a reversible "standard model" that settles all the constructive issues without any supplementary elements. However, the company also manufactures corner pieces for specific purposes. Moreover, the formworks are left "uncoupled" for on-site assembly, which reduces the encumbrance and impact of transport.

The forms are 52.5 cm tall. This vertical dimension was determined after careful evaluations of buildings standard height (interflow height, openings configuration). Being the formwork height a multiple of the building height, in controlled design conditions material waste is minimized. Furthermore, the minimum interlock modularity for each formwork amounts to 25 mm so that it can be also cut lengthwise quickly and neatly enough.

These formworks have larger dimensions than the usual systems available on the marketplace. Also, their internal structure is characterized by a double "spacer" which makes the pouring in place faster and the rebar placing easier. All finishing are made easily applicable thanks to some pre-existing characteristics of the form itself - such as the external surface, which is purposely predisposed for cement mortar and finishing resins to adhere.

The "spacers" are geometrically designed in full compliance with the current technical regulations for concrete covers and rebar spacing. The spacer also guarantees a sufficient rebars slot without further on-site fastening. It also allows the placing of thicker rebars or brackets.

Fig. 1.1 - ICF ITALIA formwork

Every ICF manufacturer provides the specific thermohygrometric values related to the thickness of their formsworks. This simplifies the designers' job. In the specific case of ICF ITALIA - taken as reference for the aforementioned reasons -, all data is listed out in the respective tables available at www.icfitalia.eu.

By way of example, transmittance can range from a maximum of 0.213 W/mq*K as for the 7.5/15/7.5 formwork to a minimum of 0.109 W/mq*K for the 15/25/15 formwork. Further data is available on the company website.

Every possible design situation has already been tested out and reported. By virtue of this, technicians can now face any design case problems. Two final drawings of an ICF retaining wall and an on-grade one follow.

Further drawings or design samples are available on the www.icfitalia.eu or www.icfpro.it websites as well as in Chapter 2.

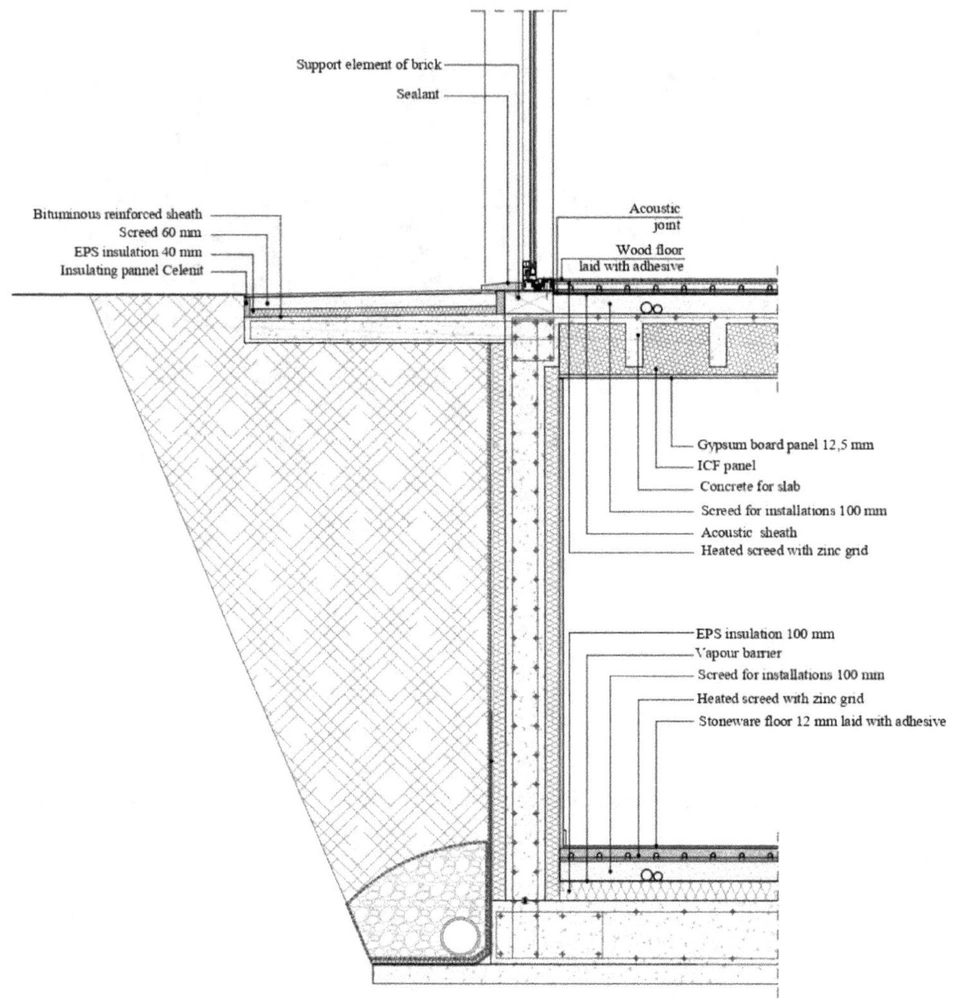

Fig. 1.2 - Drawing sample of the ICF wall of an under-grade basement. The stratigraphies and the slab attachment mode to the ground floor are shown.

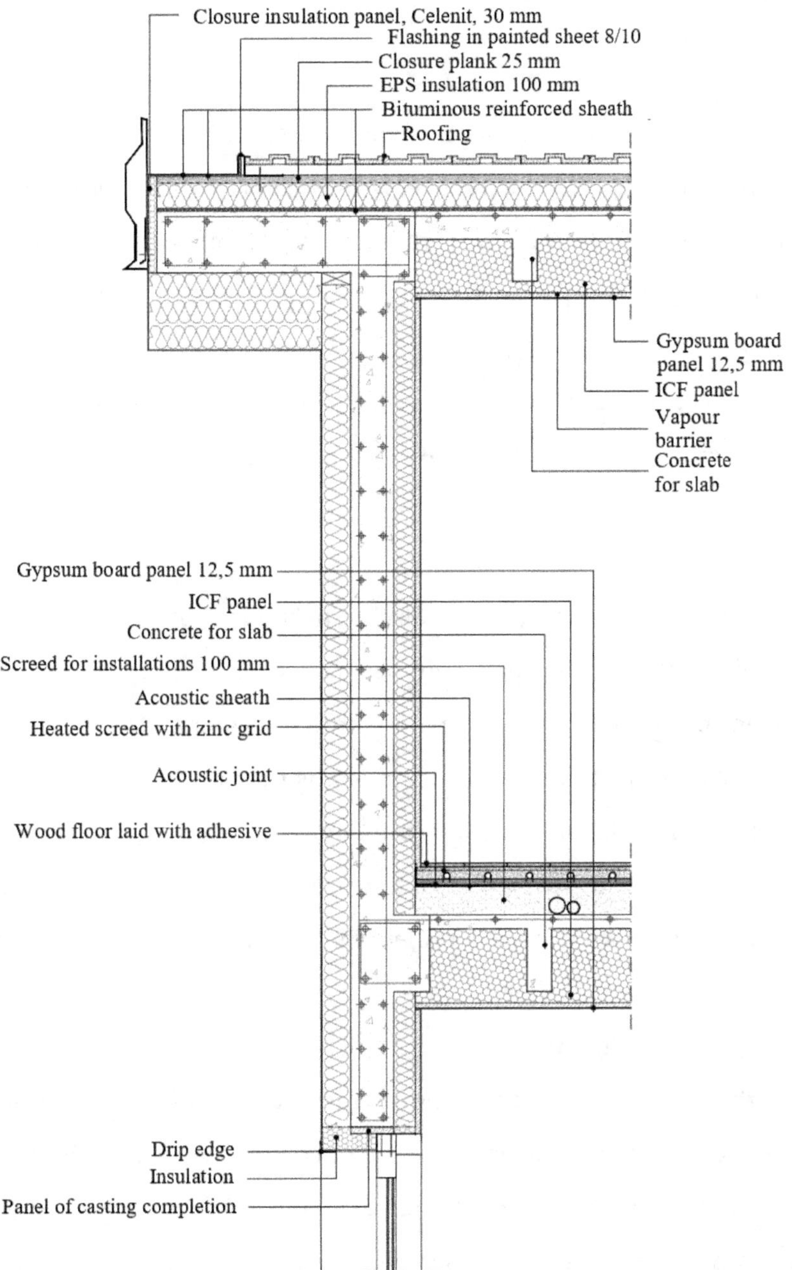

Fig. 1.3 - Drawing sample of the on-grade part of an ICF wall. The junction mode with the interflow slab and the covering is shown.

1.2)   Benefits

As the present study will show, all the collected scientific contributions agree on the major benefits the ICF constructive system has to offer to both builders and occupants. It was namely this consideration that led most of the Authors of these works to investigate the peculiar aspects of this system through testing and experimentation.

The present paragraph will describe some of these advantages concerning the design and construction phases as well as to the building occupancy.

Design Flexibility

The poured-in-place walls system can be fully integrated with conventional structural elements in steel or reinforced concrete such as isolated columns. Therefore, in specific irregular architecture designs requiring free spaces, columns can be integrated just as easily as in conventional framings. Similarly, no prescription restricts the amount and configuration of openings or the number of storys, since each building undergoes a targeted structural evaluation. As a result, the architect is "free" to conceive a project that meets the needs of their client.

Time-saving construction phase

ICFs make the construction process more time-saving. This advantage plays a key role in the entire on-site construction phase since it allows to make back sooner on the investments and therefore reduce the construction costs. Faster on-site construction also assumes paramount importance when it comes to emergency reconstruction. In such cases, a time-efficient on-site process proves to be vital. Generally speaking, ICFs bring considerable benefits to all building contractors who aim to improve the construction phase time to tender for public service works.

Sustainable building

The ICF formworks, along with the polystyrene therein, exemplifies the concept of sustainable building, since ICFs can drastically reduce energy consumption, CO2 emissions and the company energetic dependence on non-renewable sources.

The energy saving of ICF buildings is guaranteed by virtue of their highly efficient EPS shell that does not require further layering or excessive external energy supply. Furthermore, given the adequate disassembly and shattering processes, the walls are fully recyclable.

Energy saving and indoor comfort

Since their very structure is composed of a double insulating layer, ICF systems allow the construction of highly energetic efficient or even passive buildings. This prevents extra costs usually involved in traditional building. Also, the minimum transmittance values of walls and slabs, and the continuous insulating layer over the main wall junctions, prevent the occurrence of thermal bridges.

Unlike all constructive technologies involving insulation as an added element, ICFs take full advantage of the pre-existing insulating material. This minimizes the ratio between the extra construction costs for the insulation and the derived energy saving.

Cost-effective construction phase

Besides the more efficient construction methods, ICFs also guarantee on-site cost savings. This is achieved by simultaneously developing different constructive phases (structural phase, perimeter closure, insulation), by optimizing the structural materials (ex. in foundation) and by simplifying on-site construction. As this clearly shows, in the construction phase ICF systems save money compared to traditional building, especially conventional wooden constructive systems. Both direct (materials, time consumption, manpower, tools) and indirect construction costs (management, organization, interests, amortization, fixed costs for construction site) are therefore cut. This leads to a potential increase in the builder's profit and/or to the building sales price reduction.

Structural safety

Along with energy saving, the structural safety of ICFs is the strong point of the system. This comes without involving extra costs and rather with a few cost-saving factors (such as cost-effective foundations). ICF buildings can reach the highest safety results thanks to their stiffness and the low mechanical strain of the used materials (steel and concrete). The walls being properly distributed, ICFs can bring to the construction of multi-story buildings with verifiable flexibility all over the most seismic areas in Italy. The proper walls distribution can also lead to the elimination of punctiform structural elements such as columns, since the seismic response and resistance depend on bidimensional extraresistant elements. For this same reason, the amount of beams can also be reduced to the bare minimum with respect to the hierarchy of resistances. This also allows slabs to weight down directly on the walls, which consequently avoids deflection load on the floor and foundation beams.

Seismic ipersatety

Double-walled reinforced concrete load-bearing walls are characterized by plenty of resistances and limited deflection of a lower order of magnitude than that of framed walls. The designer of a building with properly distributed walls can therefore easily perform the evaluations for elasticity and flexibility as required by the current regulations (application of the full seismic spectrum) and go even farther than that. As far as civil buildings are concerned, by merely taking the full advantage of the structural materials, the building designer can afford to improve the project seismic parameters. Shortly, it is possible to design houses that resist the heaviest seismic acceleration, prevent damage, and remain operational after a severe earthquake. This is substantial since guarantees that the occupants will not need to temporarily or permanently abandon their houses even after a earthquake of the highest intensity.

Structural integrity under exceptional actions

In certain areas spread around the world, buildings need to be designed according to a structural integrity that can resist different actions other than the static and seismic ones. In the U.S. for example, along the coastal areas there is need for tornado resistant buildings whose walls cannot be penetrated by objects coming forward at maximal rate due to the wind loads (likened to big bullets). Wooden-framed buildings cannot comply with such requirements. On the contrary, ICF buildings conform to these rules without any extra costs.

On-site safety

The use of ICFs reduces the on-site risk exposure of workers. Lightweight materials to move, the reduction of the overall on-site working hours, the grouping together of multiple stages and the simplified construction process are only a few of the many benefits that make ICF sites safer than the conventional ones. The reduction of direct risks and risk interference also goes along with the lower "safety costs". This positively impacts on the on-site economy.

Extra safeguard

Nowadays, the building industry has to carefully observe the vast bureaucracy and the countless regulations for this field. This leads the professionals to be highly concerned with the repercussions their work may bring along in the long term. By simplifying the construction process and standardizing the design, ICFs minimize the chance of on-site mistakes. This also applies to the most difficult fields to handle like acoustics and thermal bridges. Moreover, ICF systems make it easier to build highly durable products whose performances do not get any lower with time. Both the builder and the final users will therefore be more safeguarded since the risk incidence and the premium on posthumous insurances will be reduced.

Increase in saleable area

ICF systems allow to build walls with an overall thickness amounting to about 30 cm while fully complying with the current structural, thermal and acoustical regulations. On the contrary, traditional construction methods can reach the same performance only with far thicker wall packs (ex. brick + thermal coat) amounting to 40-45 cm depending on the climatic area.

For example, given a cautiously estimated 10cm difference in thickness between ICF and traditional walls, and the same external dimensions, two respective regular-shaped buildings with a 1000mq footprint will show a floor area difference of 5mq per floor! Given an average market value, a builder can benefit greatly from this ICF property.

1.3)   Poured in place concrete

From the point of view of construction, it is worth pointing out that the formworks are marketed as uncoupled modular elements to optimize the transport costs. They are assembled on-site, then the integrative rebars are placed, and lastly the concrete is poured.

The rebar typology is determined by the structural designer according to the stresses and shakes the buildings will need to absorb.

The practical usage of ICF forms can be optimized by every builder within compliance with the construction specifications listed in the publications in the field (references available at Chapter 2). The following pictures exemplify a few specifications of the pouring in place, since the present volume cannot offer a complete representation of each step of the construction phase. The systems manufacturers can guarantee full technical support during both the design and on-site construction phase for any operator in need of proper training.

The reference size of the ICF ITALIA formworks is 120x52cm. They are available with 6, 7.5, 10, 15cm thick insulating pieces and 15, 20, 25, 30cm large space for concrete. This way, the formworks attempt to meet every possible structural and energetic need. In a regular building (3-5 storys), the concrete usually fills a 20cm large space. The insulation thickness varies with the climatic area, although the external shell usually measures more than the internal one with a thickness of about 7.5cm.

The particular interlocks on the superior and inferior surface of the formwork prevents the concrete from leaking out and guarantees stability during the pouring. Through easy assembly steps, ICF formworks allow to obtain already insulated reinforced concrete walls with transmittance values ranging from 0.130 to 0.235

w/mqk depending on the thickness, with thermal lag between 10 and 18 hours, and attenuation amounting to less than 0.030.

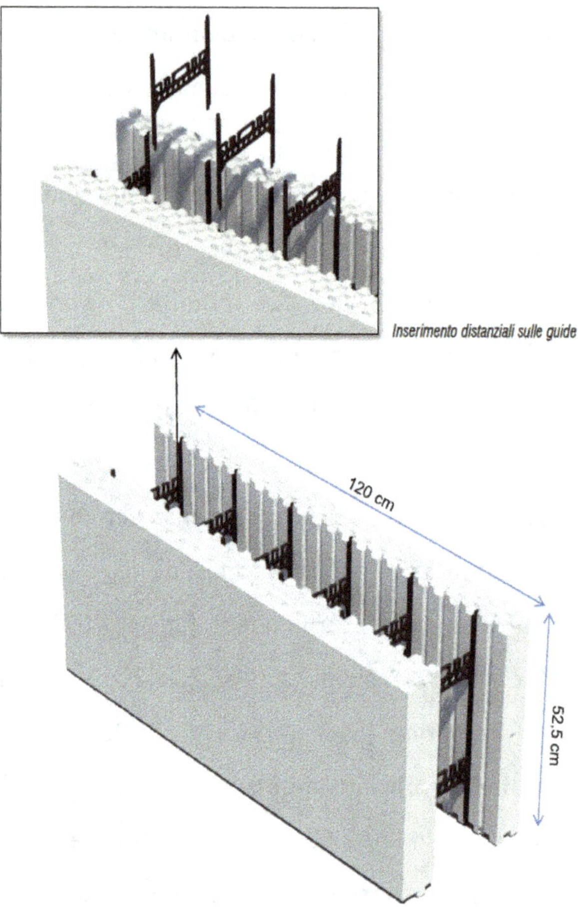

Fig. 1.4 - Scheme visual of spacers in ICF forms

Therefore, it is possible to build reinforced concrete walls cast on-site with variable thickness (15, 20, 25, 30cm) by simply placing the respective spacers onto the ICF formwork rails. By virtue of this perfectly flexible on-site pouring stage, the traditional problems related to on-site storage and management are easily solved.

Fig. 1.5 - ICF forms during assembly with some of the spacers already in place.

Small forms ICF systems are specifically meant for self-bearing walls. It is also possible to employ extendable and adjustable bracing props that once fixed to the ground brace the wall against wind overturns, this way guaranteeing the perfect sealing of the wall itself.

The props can be easily adjusted vertically through a manual vice screw, whereas a common wooden screw can be used to fix the ICF forms to the struts. The advisable spacing amounts to 120-150 cm, although it should be thickened for specific usage. Otherwise, it is also possible to brace the walls by using wooden listels fixed to the ground. They fulfil the same role as the prefabricated steel props and are easily available to every building company.

Fig. 1.6 - On-site storage of ICF forms during assembly stage. A few vertical props for walls stealing can be seen.

If the height of the wall in progress does not equal a multiple of the standard form height (cm 52.5*n), forms can be cut horizontally with a circular table saw or a hacksaw.

If the length of the form should be resized according to the wall length, the forms can as well be cut vertically along the cutting marks with a hacksaw or a cutter.

Fig. 1.7 - Worker during the assembly stage of ICF forms

The iron rebars are then placed into the formworks along the proper rails on the spacers. The horizontal bars are placed as you proceed with the pouring, whereas the vertical ones are put down from a height once the formworks have reached the full height of that story (typically 3m). By using "uncoupled" formworks as those of ICF ITALIA, the placing of more complex rebars or wall-thick beams is extremely easier - which cannot be said about the pre-assembled ICF formworks. This also allows to build complex and irregular-shaped structures.

Once the rebars are placed, the formworks verticality is verified and the closure lockings over the openings are checked, the concrete can be poured.

The concrete in-place pouring follows the same procedure as traditional construction methods. ICF systems allow the pouring in place from a height of 3-4m with nothing more than a common propping up and without any risk of deflection or failure. Having taken the proper precautions, concrete might as well be poured from greater heights.

Fig. 1.8 - Placing of a curb rebars on top of an ICF wall

The poured concrete must be fluid (slump S4 or similar). Unless otherwise indicated by the site manager, it is appropriate to slightly vibrate the concrete through either narrow hydraulic needles or a plate concrete vibrator.

Having taken the necessary safety precautions for the cast operators, a joint should be used at the tip of the pump in order to slow down the pouring concrete.

It is good practice to move the pump around the building so ast to cast the concrete in 50-60cm tall layers. This allows a slight curing of the concrete and consequently reduces the buoyancy force against the formworks.

30

Fig. 1.9 - Partially cast ICF wall

Fig. 1.10 - Cast of an ICF slab and the perimeter curb

Fig. 1.11 - ICF wall after casting (visible EPS/concrete/EPS stratigraphy)

With the ICF system, all sorts of roofing (pitched roof or flat slab) can be made.

As far as wooden roofings are concerned, the beams can be directly anchored during the casting or installed onto the walls through steel connections (clamps, grouting, and so on). As an alternative, the wooden beams can be put on top of the wall, and further form blocks or brick panels properly insulated on the outside can then close the voids.

Wooden roofings are no doubt of aesthetic interest, although from the point of view of construction, they represent a structural and thermoacoustic "discontinuity". Therefore, in order to guarantee more adequate results, the solid and uniformly insulated ICF slabs with no thermal bridges should be preferable.

Fig. 1.12 - Unfinished ICF building with wooden roofing

Before installations are positioned, "marks" must be created on the insulation with so-called "hot-blade" tools. These tools remove polystyrene pieces to create the adequate voids for the piping and the electrical junction boxes that are subsequently fixed with expansive foam.

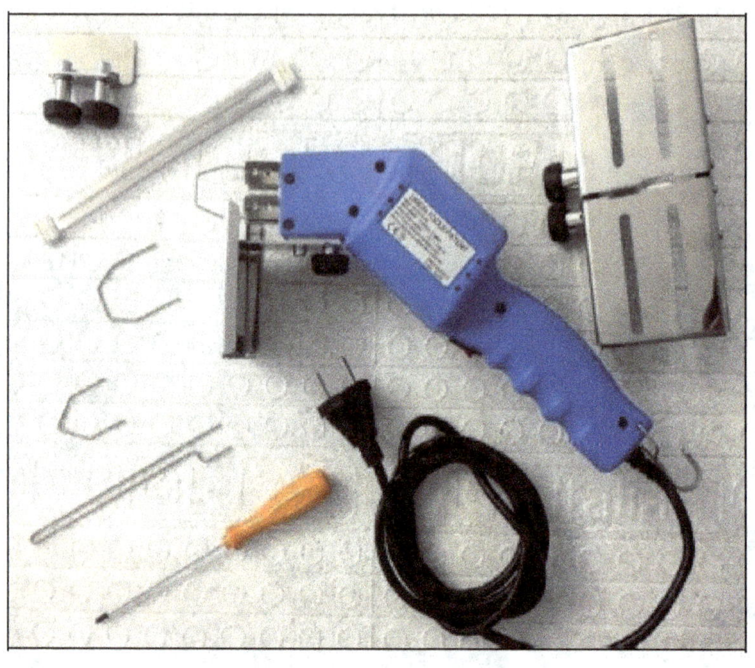

Fig. 1.13 - Hot-blade tool used to mark the ICF walls

The internal finishing of ICF walls usually requires the use of drywall sheets to be put in place after the installations have been created. These sheets can be easily and quickly installed to the formwork through mechanical fastening without further steel substructures by taking advantage of the material predispositions for direct wall screwing.

Wherever necessary, traditional plastering can also be used. However, drywall and fiberboard sheets significantly improve the walls and slabs acoustic behavior. With the specific system of reference of this study, on-site measurements showed that a 20 cm thick ICF ITALIA wall finished with a double drywall 12.5 mm thick sheet guarantees a 58dB noise reduction.

Fig. 1.14 - Interior of a ICF building with drywall sheets being applied

As far as the exterior finish is concerned, the common procedure does not differ from the one usually followed for thermal coat smoothing - that is to say, thin layers of glue and resin alternated with the adequate plastic nets.

Otherwise, it is possible to coat the building walls with decorative tile applied on the forms with the adequate glue. Different coatings either in brick, wood or steel panels can also be used. These elements must be installed on the adequate understructure that is fastened to the reinforced concrete wall. In such a way, it is possible to create ventilated facades as well.

Fig. 1.15 - ICF building coated with exposed brickwork

Fig. 1.16 - ICF building with smoothed exterior finishes

1.4) Construction designs

The proper application of the ICF constructive system and its cost and material optimization cannot prescind from an equally proper construction design. This is especially true for complex designs and buildings located in seismically prone areas. Through the adequate calculation tools, structural designers can accurately distribute load-bearing walls with proper thickness and reinforcement in full compliance with the current regulations, also taking into account the typical constructive practices of the given system.

It is thus possible to guarantee the full safety and cost-effectiveness of the building. The present paragraph will analyze a four-story house project taken as an example. This building is an on-grade apartment house with a basement located in Rimini (Italy), only a few meters away from the Adriatic coast. The area has a medium-high seismic risk (ag. max 0.238g).

Architect Luca Edera was the architectural designer and the Author of the present book, eng. Cristian Angeli, was the structural designer.

According to the client's requirements, the house was entirely designed and developed with the ICF constructive system (specifically, ICF ITALIA formworks were used) for both walls and slabs. Due to the presence of superficial water, traditional reinforced concrete was used to build the basement.

Since the client was forward-looking enough to ask for a passive building, the on-grade walls were built using 7.5/15/15 thick ICF formworks - and therefore with a 150mm thick outer EPS layer.

A careful structural modeling of the finite elements was created through the SISMICAD software (linear elastic model), produced by Concrete ltd. (Italy). Despite the building height and the seismicity of the area, the structure was solved by the structural designer with barely 150mm thick load-bearing walls (concrete core) and a small amount of reinforcement - as shown in the following pictures. Structural evaluation was carried out in full compliance with the technical Italian regulations and Eurocodes.

The on-site construction works were undertaken by a local builder. Despite this being their first experience with ICFs, the company successfully cast the entire raw structure - including the reinforced concrete basement - within 4 months. The builder stated they were fully satisfied with the progress of their work.

By the date of the present publication (March 2020), installations and interior/exterior finishes are being implemented.

A few views of the architectural project are provided as well as drawings from the structural project and the calculation model.

Fig. 1.17 – View of a lateral side of the building. The drawing was taken from the architectural plan

Fig. 1.18 – 3D view from the architectural plan

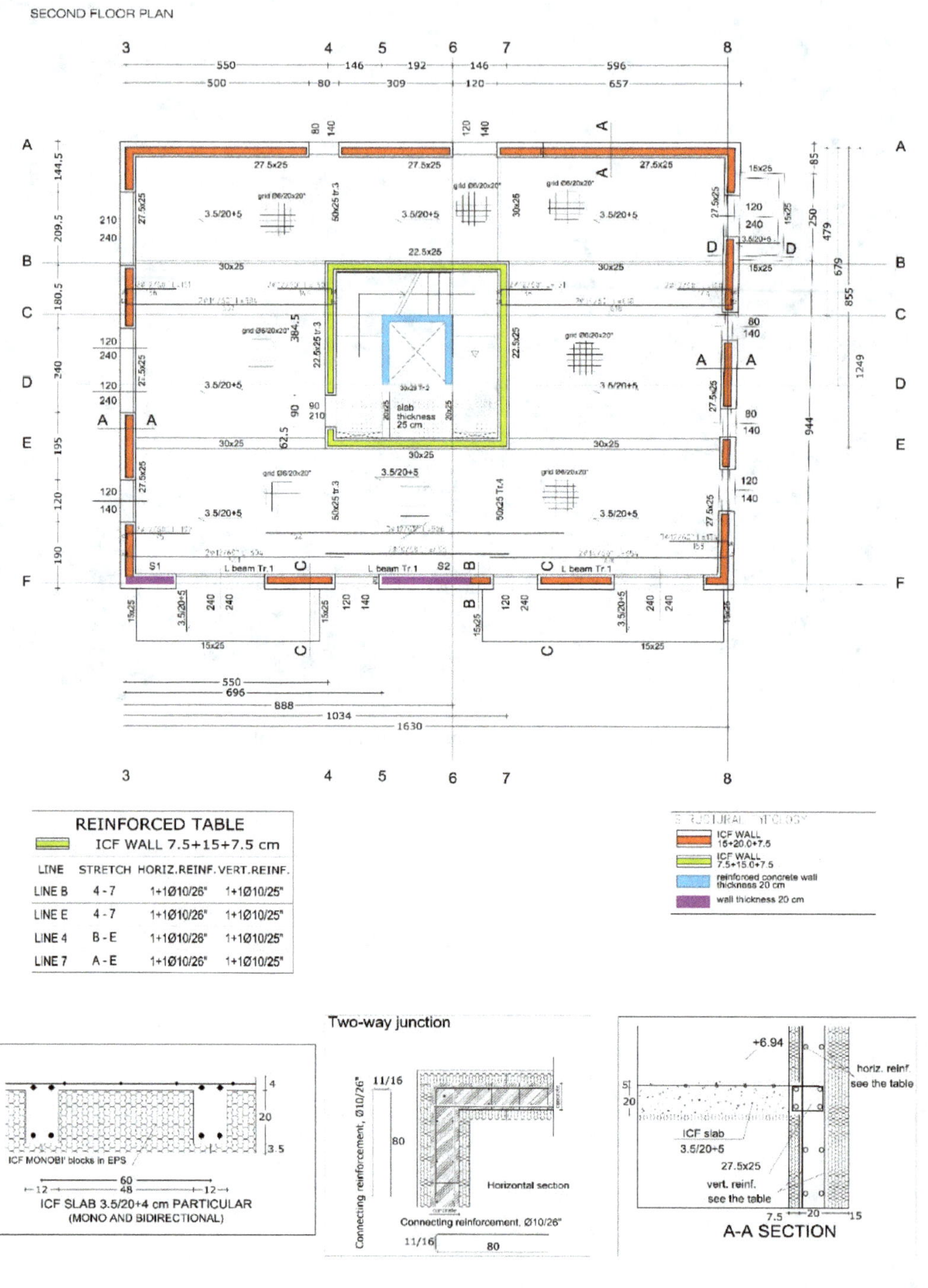

Fig. 1.19 – Structural plant of the floor type showing a few constructive details

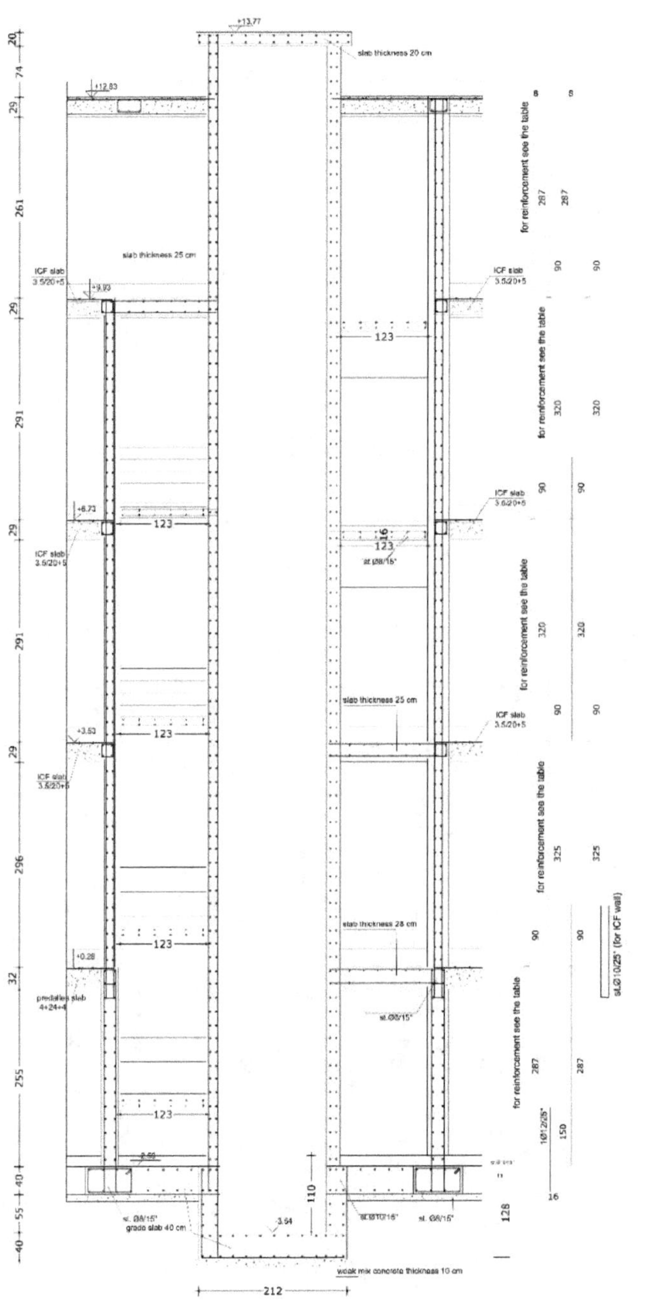

Fig. 1.20 - Structural section of the lift shaft

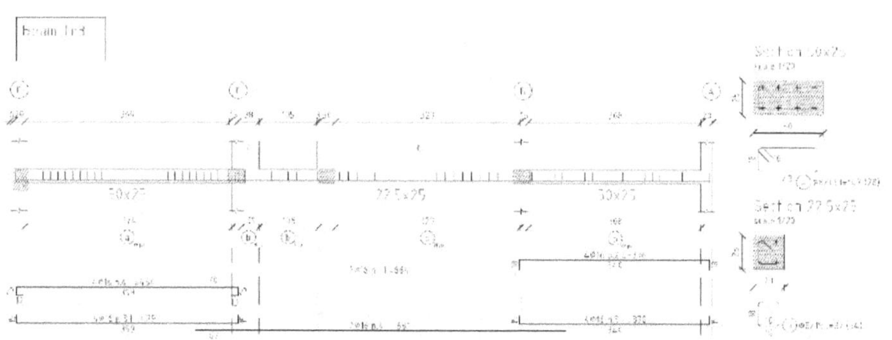

B-B SECTION

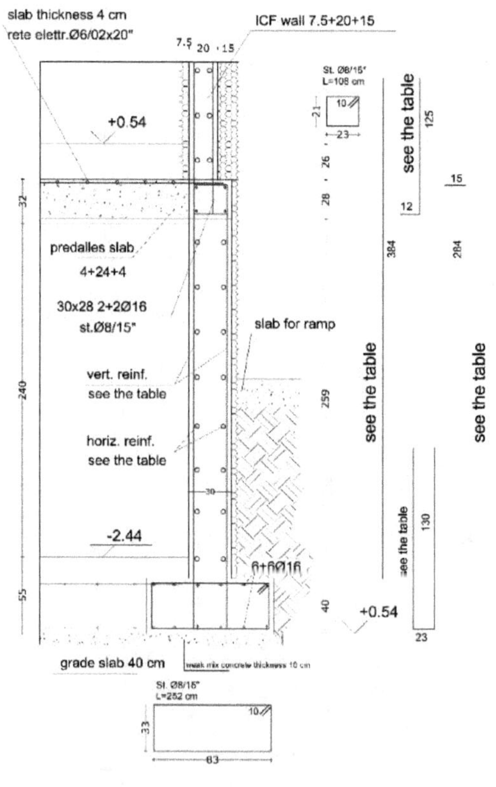

Fig. 1.21 – Detail of a retaining beam

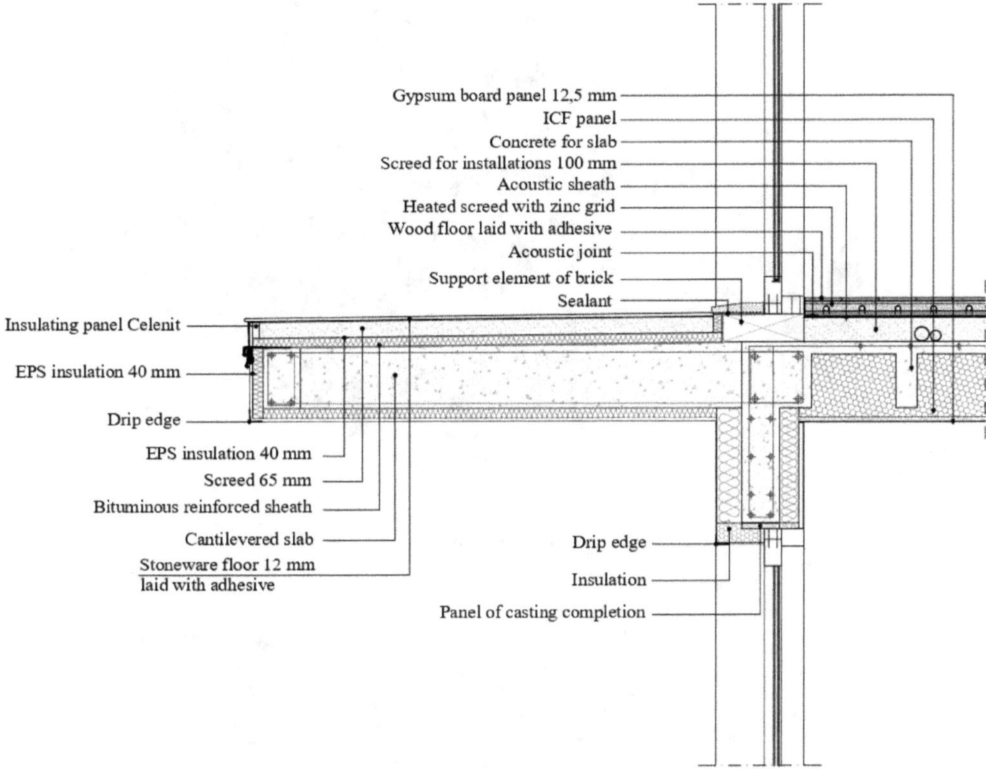

Fig. 1.22 – Detail of the balcony junction to the perimeter walls

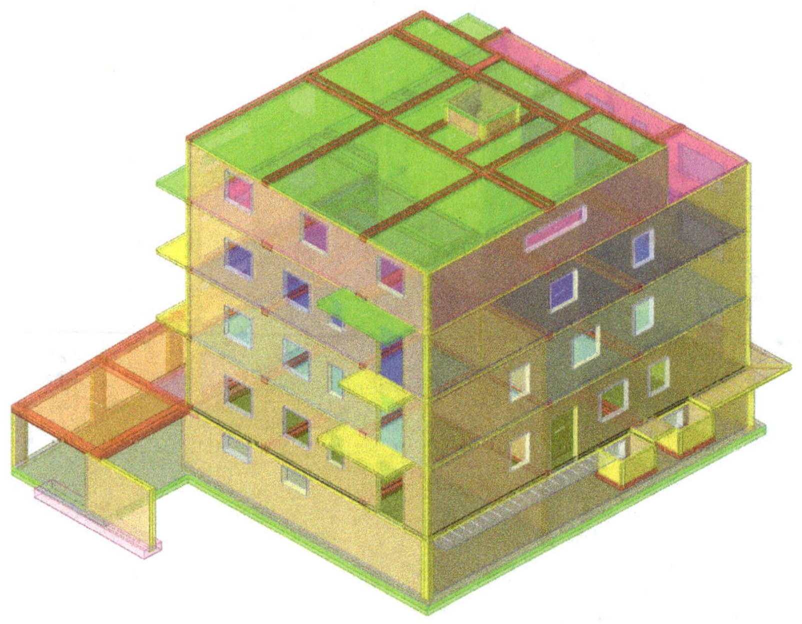

Fig. 1.23 – 3D view from the Sismicad software

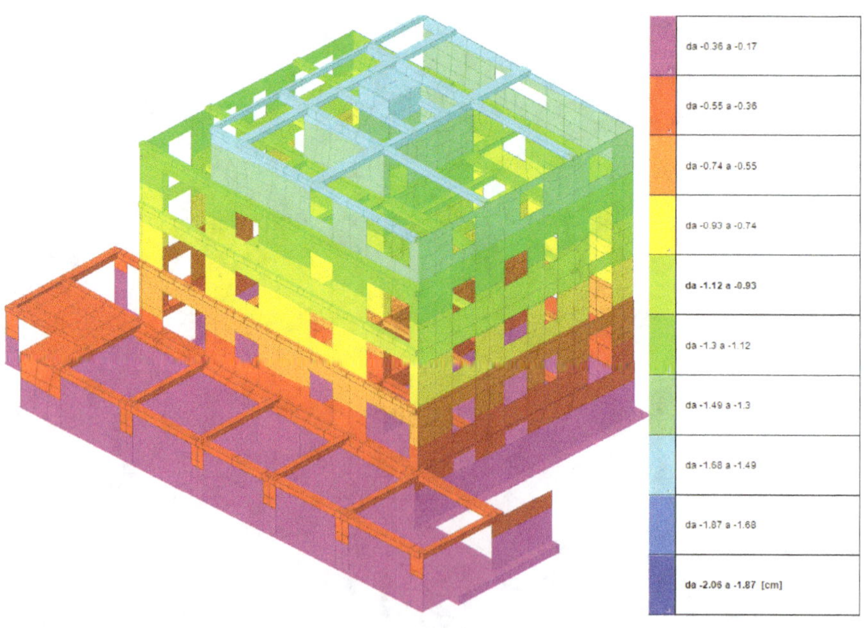

Fig. 1.24 – View of the finite element model. Maximal displacements under SLV are shown

1.5) Buildings and sites

Here follows a series of pictures illustrating the on-site construction process. Through these photographs, it is possible to understand the peculiarity and potential of ICFs as far as speed of construction is concerned. All pictures portrait ICF ITALIA system sites all over Italy, France, Portugal and Israel.

Fig. 1.25 - Site under the construction of a four-family building

Fig. 1.26 - Site under the construction of a building with basement

Fig. 1.27 - Bird's-eye view of a site under the construction of the second floor

Fig. 1.28 - Casting of the summit curbs of ICF walls by pump

The following pictures illustrate a series of finished ICF buildings to show their aesthetic and expressive potential. It is worth noting that the aesthetic value is not different than that of traditional buildings. Also, this exemplifies how ICFs can be used to build any kind of architectural design ranging from the classical style to the modern one.

Fig. 1.29 - Classic-style villa with ICF pool

Fig. 1.30 - Country-house style villa with ICF walls coated in exposed stone and superior wooden structure

Fig. 1.31 - Building with exposed brickwork coated walls (gym)

Fig. 1.32 - Interior of the previous building (wooden large span roofing)

Fig. 1.33 - Industrial building with ICF walls coated in characteristic metal panels

Fig. 1.34 - Interior of the previous industrial building

CHAPTER 2

General publications on ICF systems

This chapter collects six works presenting a general overview on ICF systems. These specific books are no scientific pieces of research, but rather studies in which the ICF technology, use and benefits are described.

The one goal of the present chapter is that to provide information on the existence of these books as well as to outline the main points illustrated in each works. For more in-depth information, please refer to the original work.

Here follows the list of the examined books in order of use:

1. **Insulating Concrete Forms construction: demand, evaluation, & technical practice (ed. 1997)**

   Prepared for: US Department of Housingand Urban Development Office of Policy Development and Research Washington, DC and The Portland Cement Association Skokie, Illinois, USA

   Prepared by: NAHB Research Center, Inc. Upper Marlboro, Maryland, USA

2. **Prescriptive method for Insulating Concrete Forms in residential construction (ed. 1998)**

   Prepared for: US Department of Housingand Urban Development Office of Policy Development and Research Washington, DC and The Portland Cement Association Skokie, Illinois, USA

   Prepared by: NAHB Research Center, Inc. Upper Marlboro, Maryland, USA

3. **Costs and benefits of insulating concrete forms for residential construction (ed. 2001)**

   Prepared for: US Department of Housingand Urban Development Office of Policy Development and Research Washington, DC and The Portland Cement Association Skokie, Illinois, USA

   Prepared by: NAHB Research Center, Inc. Upper Marlboro, Maryland, USA

4. **Use of Insulating Concrete Forms in residential housing construction (ed. 2000)**

   Author Lewis, Dan C., Florida, USA

5. **Design and construction using Insulating Concrete Formwork**

   Author A.K.Tovey, J.J. Roberts, M. Kilcommons, Concrete Centre, USA

6. **Sistemi costruttivi a pareti portanti in c.a. Insulating Concrete Form**

   Author C. Angeli, Legislazione tecnica, 2018, Italy

7. **Edifici sismoresistenti progettati con le NTC 2018. Analisi e comparazione tecnico economica tra sistemi costruttivi**

   Author C. Angeli, Legislazione tecnica, 2018, Italy

In the following section, a short description of each work will be presented.

From the review of the aforementioned studies (with specific reference to first and fifth books listed), it appeared that the current American and Anglosaxon design criteria for ICFs are much simpler than those required by the European regulations. As a result, a discrepancy in the analytical and experimental approach between the two is exhibited by the scientific reports in Chapter 3.

1. **Insulating Concrete Forms construction: demand, evaluation, & technical practice (ed. 1997)**

Prepared for: US Department of Housingand Urban Development Office of Policy Development and Research Washington, DC and The Portland Cement Association Skokie, Illinois

Prepared by: NAHB Research Center, Inc.Upper Marlboro, Maryland

This book was written and published in the U.S. for the Portland Cement Association. It concerns the American methodologies and executive standards. The book is divided into 3 sections and an appendix.

The first section is dedicated to a broad overview on ICF systems and their related American codes. It also gives a comprehensive description on the ICF thermal and structural behavior as well as the construction process.

The second section is the key of the whole publication and outlines four different ICF buildings all over the U.S. (Virginia, Texas, Iowa, and Maryland) also providing a number of pictures of the process itself. The description does not simply cover the executive aspects, but it also gives each buildings design, marketing and costs analysis. All of these factors are reviewed by both the builder and the buyer - which provides multiple standpoints on each topic. In order to present a reliable and impartial work, the four projects were all built using a different ICF system (grid or flat walls) that is accurately described at the beginning of the related paragraph.

The third section summarizes the conclusions drawn from the first two sections.

Lastly, the appendix gives a few concise cost evaluations starting by the analytical ones presented in each section. It is full of comparative tables and technical data.

Fig. 2.1 - Frontispiece of the volume titled "Insulating Concrete Forms Construction: Demand, Evaluation, & Technical Practice"

2. Prescriptive method for Insulating Concrete Forms in residential construction (ed. 1998)

Prepared for: US Department of Housingand Urban Development Office of Policy Development and Research Washington, DC and The Portland Cement Association Skokie, Illinois

Prepared by: NAHB Research Center, Inc. Upper Marlboro, Maryland

The content of this book is accurately contained in the title itself. At the beginning, a "prescriptive method" is outlined whose dimensioning is applicable to ICFs - that is to say, a set of rules taking into account the American technical regulations and constructive rules, and aiming to standardize the use of ICFs. In doing so, the Authors hope to reduce the need of resorting to design engineers for modest scale buildings.

This part is followed by two sections. The first one deals with the implementation of the forenamed prescriptive method to illustrate its use and effectiveness. The second one allows the dimensioning based on the typology and size of the building. It is worth pointing out that the prescriptions examined in the volume only apply to regular buildings (ex. detached or semi-detached terraced houses). For complex constructions or structures under exceptional loadings such as buildings along the coastal areas prone to hurricanes, different rules apply.

It is particularly interesting to note a table indicating the dimensional restrictions for structural elements (foundations, walls, slabs and so on) to which the prescriptions apply - that is to say, which do not need any customized design.

Oddly enough, however, this "prescriptive method" considers the internal walls reinforcement to be a single and optional rebar that should be placed into the wall midpoint only in specific cases.

Lastly, the Prescriptive method for insulating concrete forms in residential construction states that reinforcement is not needed along constructive interconnections such as that between foundation and wall. According to the Authors, the reinforcement is only required whenever the backfill height exceeds 4 feet (1.2m), and in areas with higher wind speed than 130 mph (209 km/hr) or located in D1 and D2 seismic areas.

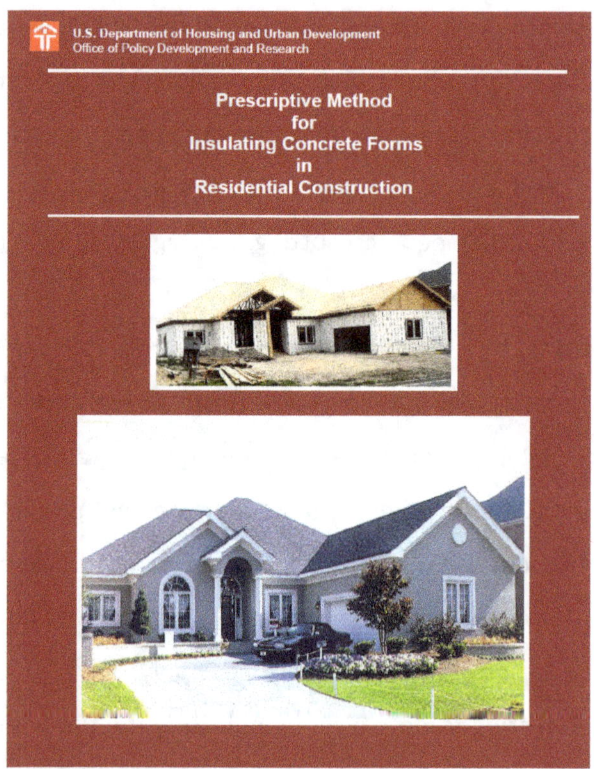

Fig. 2.2 - Cover of the book titled "Prescriptive method for Insulating Concrete Forms in residential construction"

3. **Costs and benefits of Insulating Concrete Forms for residential construction (ed. 2001)**

Prepared for: US Department of Housingand Urban Development Office of Policy Development and Research Washington, DC and The Portland Cement Association Skokie, Illinois

Prepared by: NAHB Research Center, Inc. Upper Marlboro, Maryland

This relatively short study examines the construction costs of ICF buildings. The data comes from several sites all over the U.S. and show that in that specific area, the construction costs of an ICF building usually exceed those of a wooden equivalent by 3-5%. This means a small difference in dollars per mq. In the long term, however, the cost gap is evened out by the comfort, structural integrity, fire resistance and energetic savings of ICF buildings.

Also, this publication presents a noteworthy series of pictures of ICF buildings after a tornado. The pictures document how every coating and finish element including the roofing is taken away by the wind, whereas the ICF walls keep stable.

By reporting the qualitative considerations drawn on the basis of the various performance standards, the NAHB Research Center assesses through a number of case studies the higher performance of ICFs over traditional wooden framing.

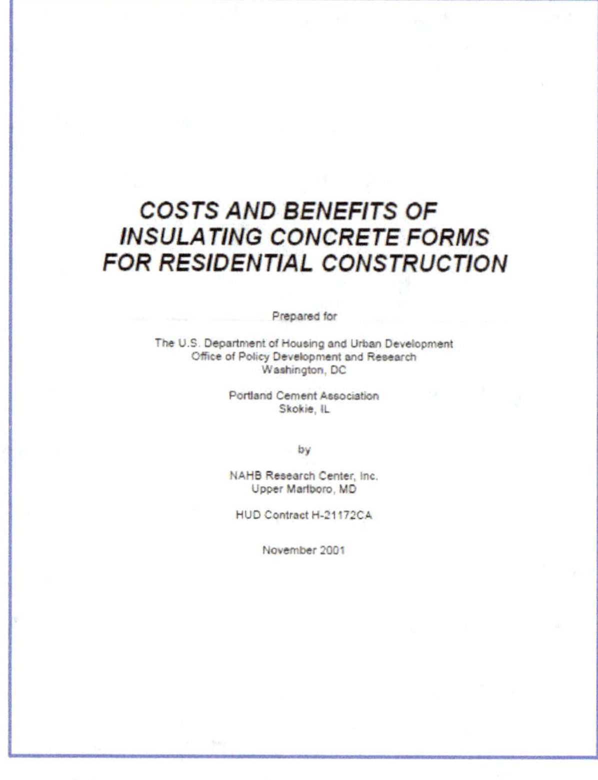

Fig. 2.3 - Frontispiece of the volume titles "Costs and benefits of Insulating concrete forms for residential construction"

4. Use of Insulating Concrete Forms in residential housing construction (ed. 2000)

Author Lewis, Dan C., University of Florida

This work starts by briefly listing out the various benefits of ICFs (structural, thermal and fire properties, termite protection, and so on). It also lists the commercial typologies of ICF formworks with specific reference to American ICF houses. Each typology is documented with pictures and is composed of a different insulator (EPS, XPS, polyurethane and aggregations) whose mechanical and chemical properties are carefully described.

This part is followed by a comparative study between the costs and performances of an ICF building and a traditional wooden framing.

This comparative study analyzes the construction processes, the costs, the timing, and the performances (sound insulation, durability, fire resistance) of both buildings. As far as the costs are concerned, the Author points out through tables and graphs that in the U.S., ICFs tend to cost slightly more than the traditional wooden framing (2-4%). However, this gap is fully offset by the higher energetic, acoustic, and fire resistant performances of the ICF building.

The book is full of graphs comparing by performance the wooden framing with the ICFs. The Author highlights the advantages of ICFs especially on the performative elevel of fire resistance, since "fire wall tests, where the walls were heated with gas flames at 2000 degrees Fahrenheit for 4 hours, were conducted to compare ICF and wood walls. Wood frame walls earned a fire rating range from 0.75 to 1.0 hours, while the fire rating for ICF walls ranged from 2.0 to 4.0 hours".

Lastly, the work offers a few interesting ICF drawings by typology (panel, plank, block) and by load-bearing internal structure configuration (flat panel, waffle-grid, post, and beam).

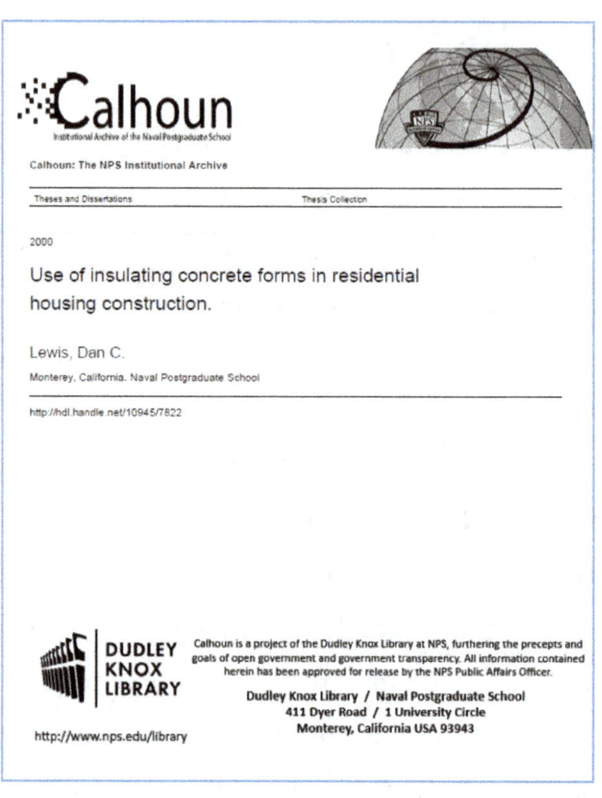

Fig. 2.4 - Cover of the volume titled "Use of Insulating Concrete Forms in residential housing construction"

5. Design and construction using Insulating Concrete Formwork

Author A.K.Tovey, J.J. Roberts, M. Kilcommons, Concrete Centre, USA

The book reviews various ICF systems and the different structures they can be used for. Many design aspects and regulatory factors referring to the BS and Eurocodes are described.

It is worth noting that this work presents the first of its kind section offering test reports to fire resistance of ICFs with 150mm thick concrete. From the attached graphs it is possible to observe:

the temperature trend of the concrete opposite surface as a function of the distance from the flame and the length of exposure;

the temperatures of the unexposed surface (with concrete finish measuring 13mm) never exceeding room temperature (16 degrees) within 90 minutes nor 60 degrees even after 3 hours.

The minimum thickness for the ICF walls structural part indicated in Design and construction using insulating concrete formworks amounts to 120mm. Furthermore, the book does not consider reinforcement as a required element, but rather just "[...] typically used in concrete and masonry walls to increase flexural or vertical capacity".

Lastly, the study offers a series of noteworthy constructive drawings.

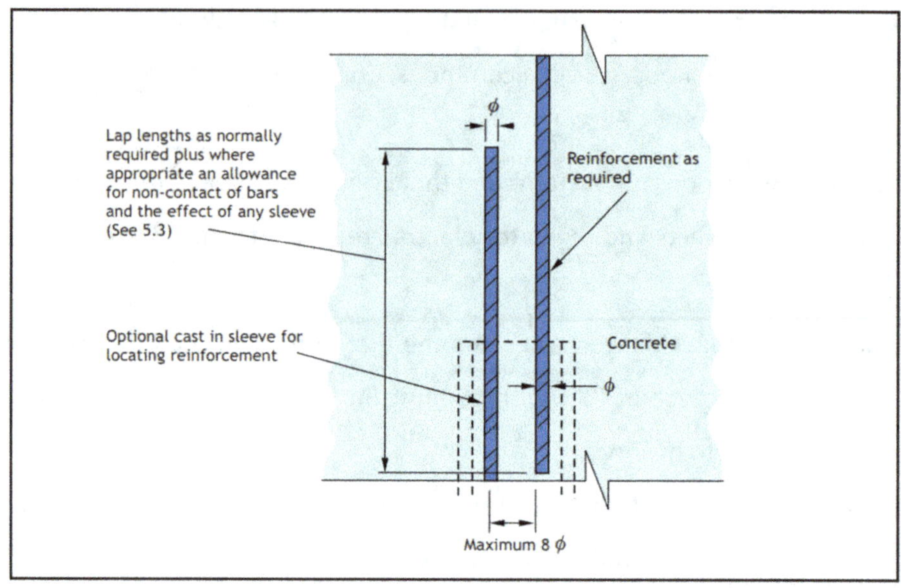

Fig. 2.5 - Rebars lap length (picture taken from the volume titled "Design and construction using Insulating Concrete Formwork")

6. Sistemi costruttivi a pareti portanti in c.a. Insulating Concrete Form

Author C. Angeli, Legislazione tecnica, 2018, Italy

Over the past years, a sequence of seismic events has taken place in Italy. The post-emergency reconstruction that followed highlighted the advantages of "innovative" construction methods combining antiseismic structures, fast on-site construction and high energetic efficiency.

In this framework, this study aims to illustrate the different reinforced concrete load-bearing wall systems available on the marketplace. Each system is described from the standpoint of technology, regulation compliance, manufacturing, design and implementation in such a way as to show their real potential.

More specifically, the book investigates all aspects of load-bearing wall ICF systems with polystyrene formworks, whose characteristics are compared to those of conventional construction methods.

Plenty of experimental data on case studies the Author was involved in is provided as well as a few "model" designs. Also, the many specific details on the ICF systems are easily usable and available also through computer support.

Thanks to this approach, the reader is carefully guided through the plenty of design and construction systems so as to find the most suitable one.

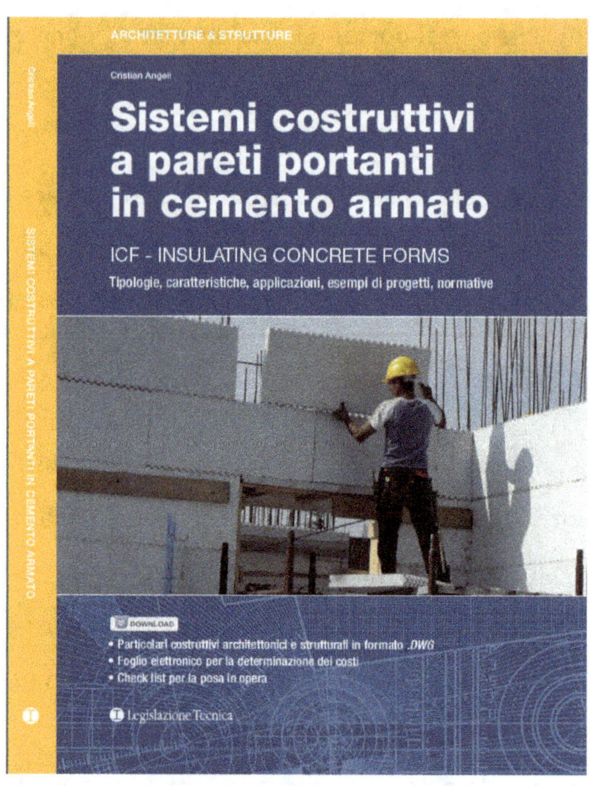

Fig. 2.6 - Cover of the volume titled "Sistemi costruttivi a pareti portanti in c.a. Insulating Concrete Form"

7. Edifici sismoresistenti progettati con le NTC 2018. Analisi e comparazione tecnico economica tra sistemi costruttivi

Author C. Angeli, Edizione Legislazione tecnica, 2018, Italy

This study provides the technical and economic analysis of an antiseismic building with three different design solutions, both traditional and innovative ones, namely: reinforced concrete framing, load-bearing ICF walls, and prefabricated wooden X-Lam forms. The three building solutions are compared by technology and analyzed by bill of quantities breakdown. In doing so, the Author hopes to enable the reader to understand what solution is the most "appropriate" given the same quality of service.

All buildings were designed in full compliance with the Italian updated technical regulations (NTCs 2018). This way, the study also offers an overview on the real changes introduced in three common use building systems from a legislative perspective. This approach is likely to apply to every European country since the forenamed NTCs are very similar to the Eurocodes.

Lastly, the book gives the first (and only) economic comparison available between these three solutions, allowing in such a way to fully understand the advantages of reinforced concrete load-bearing wall systems.

Fig. 2.7 - Cover of the volume titled "Edifici sismoresistenti progettati con le NTC 2018. Analisi e comparazione tecnico economica tra sistemi costruttivi"

CHAPTER 3

Research and experimentation on ICF systems

Contents

3.1) Mechanical and structural analyses

The present paragraph lists a series of scientific works outlining the mechanics and structure of ICF systems - both in broad terms and with a specific focus on the materials employed. Ten studies were selected. The paragraph provides the summary of each of them. These summary reports, however, do not intend to replace the original studies, that are easily accessible through the plenty of web references.

Lastly, this paragraph is concluded with a further list of studies not namely concerning ICF walls. that were not analyzed in the present work due to lack of space.

3.1.1) Compressive strength of Insulated Concrete Form blocks

3.1.2) Inspection of properties of expanded polystyrene (EPS), compressive behaviour, bond and analytical examination

3.1.3) A comparative study between wall bearing steel reinforced expanded polystyrene composite wall system and Insulated Concrete Forms

3.1.4) Research contributions to the seismic performance of ICF technology wall systems

3.1.5) Seismic evaluation of a green building structural system: ICF grid walls

3.1.6) In-plane lateral load resistance of wall panels in residential buildings

3.1.7) Finite elements analysis for cracking control of reinforced concrete walls prone to shrinkage

3.1.8) Experimental investigation of Insulated Concrete Form (ICF) wall panels under quasi static cyclic load

3.1.9) In-plane shear resistance of Insulated Concrete Form walls

3.1.10) Characteristics of expanded polystyrene and its impact on mechanical and thermal performance of Insulated Concrete Form (ICF) system

3.1.11) Seismic-proof buildings in developing countries

3.1.12) Further references

3.1.1) Compressive strength of Insulated Concrete Form blocks

The article starts by indicating the properties and advantages of ICF systems - including fast on-site construction processes, sound insulation, energy saving and sustainability. These characteristics all meet the needs of the highly earthquake prone Indian territory (the author's country of origin).

Firstly, the article proceeds by analyzing the sample blocks. The samples are composed of two 100mm, 75mm, and 50mm thick expanded polystyrene sheets interconnected by 8mm steel rods. The cavity of the formwork is for concrete to fill. The characteristics of the EPS sheets under tested are described in detail as well as the samples curing mode. The sample formworks measured 200x150x60mm *[Ed. with a 60mm concrete layer]*.

Secondly, all specimens underwent cyclic compression tests whose results, according to the researcher, show that the sheets successfully held the concrete from falling.

Pictures and graphs illustrating the load-deflection curve are provided. These graphs show that the ductile behavior of the specimens increases with increased EPS thickness. A slight increase in the EPS thickness therefore leads to a slight increase in the specimen axial compressive strength. In this specific case, the highest values for compressive strength were exhibited by 75mm sheets.

Lastly, the article provides a table indicating the ductility ratio, which shows the highest peak namely with 75mm sheets.

SUMMARY SHEET:

Title:

Compressive strength of Insulated Concrete Form blocks

Authors:

E. Arunraj, A. Arun Solomon, G. Hemalatha

Prepared for:

Department of Civil Engineering, Karunya University, Coimbatore, Tamil Nadu, India

Publication type:

Conference proceedings

Date of publication:

April 2014

Publisher:

ISSN:

Content summary:

Axial compressive strength of ICF form blocks

Lunghezza e caratteristiche del testo:

6 folders containing pictures, drawings and graphs

URL for the full text:

https://www.researchgate.net/publication/261798817_COMPRESSIVE_STRENGTH_OF_INSULATED_CONCRETE_FORM_BLOCKS

3.1.2) Inspection of properties of expanded polystyrene (EPS), compressive behaviour, bond and analytical examination

The articles starts by defining the advantages that ICF systems can bring to the Indian building marketplace, and by carefully describing the study objectives and the ICF Indian regulations according to which it was conducted.

Firstly, the mechanical properties of the polystyrene used for the samples are investigated beforehand through laboratory testing. In order to do so, cubic sample forms were created and their axial compressive strength and deflection under loading were examined. The results are shown in diagrams indicating the load-deflection curve of the materials until failure. The data is in line with the previous literature on expanded polystyrene.

Secondly, three samples typologies are defined as follows:
- Samples casted with plain EPS;
- Samples casted with corrugated EPS;
- Samples with "naked" concrete and no EPS formworks.

Each formwork typology was tested with varying EPS thickness and density before the concrete curing. 12 samples with plain EPS, 12 specimen with corrugated EPS, and 2 concrete formworks were created. The sample formworks size was 150x200mm. Thickness varied with that of EPS.

The axial compression tests led the Authors to the following conclusions:
- The compressive strength of a sample wall composed of ICF formworks increases if the EPS is plain (since under the same thickness, the corrugation clearly reduces the amount of concrete);
- The EPS formworks influence the mechanical behavior of the samples by holding the concrete;

- The EPS formworks show higher ductility than naked concrete samples, since the EPS sheets improve the ductile behavior despite leaving the axial compressive strength unaffected.

The results allowed to derive a ductility ratio for each typology analyzed. The data shows that the formworks with 12daN/mc dense and 100mm thick EPS have the highest ductility. A table with the load and deflection values of all typologies is provided.

SUMMARY SHEET:

Title:

Inspection of properties of expanded polystyrene (eps), compressive behaviour, bond and analytical examination of Insulated Concrete Form (ICF) blocks using different densities of EPS

Authors:

A. Arun Solomon, G. Hemalatha

Prepared for:

Department of Civil Engineering, Karunya University, Coimbatore, Tamil Nadu, India

Publication type:

Scientific paper

Date of publication:

January 2017

Publisher:

International Journal of Civil Engineering and Technology (IJCIET)

ISSN:

Print: 0976-6308 and ISSN Online: 0976-6316

Content summary:

Definition of the mechanical properties of polystyrene and analysis of the polystyrene properties influence on ductility and compressive strength of ICf formworks

Text length and type:

14 folders containing pictures and graphs

URL for the full text:

https://www.researchgate.net/publication/313185325_Inspection_of_properties_of_Expanded_Polystyrene_EPS_Compressive_behaviour_bond_and_analytical_examination_of_Insulated_Concrete_Form_ICF_blocks_using_different_densities_of_EPS

3.1.3) A comparative study between wall bearing steel reinforced expanded polystyrene composite wall system and Insulated Concrete Forms

In this study, the Authors compare two concrete wall systems that both use polystyrene as an insulator, namely: the usual ICF system and the SIP (Structural Insulating Panel) system.

This comparative study specifically refers to the building marketplace of Egypt (the Authors' country of origin). The Egyptian framework is characterized by a great demand for cost-effective house units. Therefore, these constructive systems are analyzed by construction costs, speed of construction and environmental sustainability.

Firstly, the two systems are carefully described in their components, construction processes and evolution from the outset. No specific ICF system is mentioned, whereas the SIP system used is produced by the Italian manufacturer M2. The M2 technology is composed of a corrugated polystyrene core with two thin wire mesh reinforcements interconnected by wire mesh steel ties. The SIPs are encased later with two thin layers of sprayed concrete (grout), this way creating a concrete/polystyrene/concrete sandwich.

The article tends to highlight the advantages of SIPs over ICFs as far as the Egyptian marketplace is concerned. The Authors strongly stress the fact that the SIP system is so flexible it can be used as floor slabs, whereas ICFs require to install further elements.

Secondly, a few numeric evaluations on both systems are shown through pictures and graphs.

Furthermore, the article provides a table indicating that according to the Authors, the SIP system allows to build thinner walls with higher ultimate load. The comparative analysis is carried out by maintaining a constant concrete thickness (50+50mm for SIPs and 100mm for ICFs).

[Ed. This article tends to emphasize the advantages of the SIP system over ICFs. After several first-hand experimentations and construction experiences with this specific product, I feel that the final judgment should have been formed after closer examination of the advantages of SIPs as well - for example, the fact that by using the polystyrene as a structural support for concrete, the formworks cannot face problems like fires or exposure to intense heat. Moreover, as far as compressive strength is concerned, it should also be noted that thin concrete sheets of 40-50mm on both sides can hardly prevent accidental irregularities or constructive defects. It seems to me that an accurate judgement on these two constructive systems can only be formed on the basis of a complete analysis.]

SUMMARY SHEET:

Title:

A comparative study between wall bearing steel reinforced expanded polystyrene composite wall system and Insulated Concrete Forms

Authors:

Ayman El-Alfy and Ashraf Shalaby

Prepared for:

National Research Center, Cairo, Egypt

Publication type:

Scientific paper

Date of publication:

March 2011

Publisher:

82

Australian Journal of Basic and Applied Sciences

ISSN:

ISSN 1991-8178

Content summary:

Comparative analysis of two construction systems with insulated walls

Text length and type:

10 folders with drawings and graphs

URL for the full text:

https://www.researchgate.net/publication/268258465_A_Comparative _Study_between_Wall_Bearing_Steel_Reinforced_Expanded_Polystyren e_Composite_Wall_System_and_Insulated_Concrete_Forms

3.1.4) Research contributions to the seismic performance of ICF technology wall systems

This article was probably requested by the AMVIC system manufacturer, since they are frequently nominated and all ICFs showed in the pictures were manufactured by them.

Firstly, the study closely examines the properties and advantages of the ICF system and fully describes the pouring in place process.

Secondly, the article explains that the examined system allows to place the rebars in such a way as to comply with the current European seismic regulations, specifically Eurocodes. The post-elastic response induced by the confinement rebars is investigated in its compliance with these regulations as thoroughly as the different walls configurations - that although various, can be divided into two main categories on the basis of where confinement of the end zones is needed and where not.

Additionally, further seismic evaluations were carried out. The various seismic accelerations the Romanian area is prone to were tested under different concrete classes and varying wall thickness ranging from 150 to 200mm.

The test results for each configuration are shown in a curvature graph.

Lastly, the researchers end by distinguishing between "short" and "tall" buildings, which differ in the necessary shear resistance. According to the Authors, the first building type (short buildings) does not need confinement of the end zones, whereas the second building type (taller buildings) needs confinement to increase the ductility, and is therefore preferable.

3.1.5) Seismic evaluation of a green building structural system: ICF grid walls

This article introduces a different ICF formwork type from the ones described so far, namely the grid ICFs. ICF grid walls are composed of formworks whose core contains polystyrene link portions (of non-negligible dimension) interconnecting the external sheets. This means that the concrete conformation will present typically alternate full and empty spaces - from which the definition of "grid wall" is derived.

This peculiar structure differs markedly from the plain concrete wall type required by the international regulations. Therefore, it was found necessary to perform comprehensive testing of its mechanical properties.

The article describes an experimental program of cyclically loaded ICF grid walls, of big dimension (2x2m) and varying in thickness, put under shear forces until failure.

As the article emphasizes, this wall system is highly sustainable since the walls can be built out of different recycled materials other than polystyrene. However, the Authors also point out the difficulty in using this technology in high seismic hazard zones. This is due to the fact that each manufacturer produces formworks of different shapes, which leads to potential voids and multiple formwork configurations. Therefore, the mechanical behavior of each of the derived walls should be properly defined through similar testing so that an equivalence of parameters can be established with flat walls.

Additionally, the paper stresses the difficulty in reading the experimental results, since the external formworks prevent to observe how the concrete failure mechanism is triggered.

Lastly, the results of the cyclical testing are provided through plenty of load-drift graphs and the samples pictures.

SUMMARY SHEET:

Title:

Seismic evaluation of a green building structural system: ICF grid walls

Authors:

Peter Dusicka, Thomas Kay

Prepared for:

Department of Civil Engineering, Portland State University

Publication type:

Conference proceedings

Date of publication:

April 2009

Publisher:

ISSN:

Content summary:

Seismic performance of ICF grid walls

Text length and type:

4 folders with pictures and graphs

URL for the full text:

https://www.researchgate.net/publication/269122313_Seismic_Evaluation_of_a_Green_Building_Structural_System_ICF_Grid_Walls

3.1.6) In-plane lateral load resistance of wall panels in residential buildings

This long study reports the results of an in-depth experimental program in which wall panels of different types were subjected to in-plane loads comparable to those of earthquakes and wind (70mph).

The study aims to compare different formwork types by the ASTM regulations guidelines. The examined wall types are as follows:
- Wood-frame;
- Steel-frame;
- ICF flat;
- ICF screen-grid;
- ICF waffle-grid.

As the paper shows, ICF flat walls were found to be more resistant than any other wall type analyzed - their resistance is higher by a full order of magnitude than that of wood-frame and steel-frame walls.

Due to their low lateral deflection and little damage, ICF walls are therefore particularly adequate for buildings on high seismic hazard zones. This also allows to reduce the reconstruction costs.

Lastly, the study suggests a few improvements to be carried out on wood-framed and steel-framed panels in order for their stiffness to increase.

All results are provided from an analytical point of view through accurate descriptions of the testing equipment, pictures of the specimen design and graphs indicating the scientific data.

SUMMARY SHEET:

Title:

In-plane lateral load resistance of wall panels in residential buildings

Authors:

Armin B. Mehrabi

Prepared for:

Portland Cement Association

Publication type:

Research contribution

Date of publication:

2000

Publisher:

Portland Cement Association

ISSN:

Content summary:

Comparative study on the response to shear loads of different ICF wall systems

Text length and type:

77 folders containing pictures and graphs

URL for the full text:

http://citeseerx.ist.psu.edu/viewdoc/download?doi=10.1.1.1012.1173&rep=rep1&type=pdf

3.1.7) Finite elements analysis for cracking control of reinforced concrete walls prone to shrinkage

This study starts from the assumption that the issue of shrinkage in concrete is becoming increasingly serious with the spread of high-resistance concrete. Therefore, the Authors investigated this problem through the case study of a large-sized concrete building.

The work starts by defining the shrinkage phenomenon, outlining its causes, and describing its effects on concrete products. It then proceeds to describe the procedure adopted for the numerical analysis. As the Authors underline, shrinkage occurs in massive structures as often as in thinner wall systems *[Ed. and therefore, potentially in ICF buildings]*. The main goal expressed in the study is that to evaluate the structural effects of concrete shrinkage and cracking.

The study is complex and draws the following conclusions:

- Reinforcement improves the mechanical performance of the concrete structures (especially compressive strength, stiffness and ductility). However, it also produces negative side effects, such as the tendency of the shrinking walls to deflect lengthwise. This means that by opposing the shrinkage force, the rebars create an internal constraint that is likely to encourage cracking.

- Due to their crosswise position, brackets tend to make the concrete cut section "weaker". This anticipates the cracking and increases damage at the base of the wall.

- In walls prone to shrinkage, the coactions mainly occurs due to the constraints in the foundation.

- Traditional reinforcement does not prevent cracks smaller than 0.33mm. Integrated steel fibers can help with bigger deflections.

- If integrated into traditional rebars, steel fibers optimize the overall reinforcement.

SUMMARY SHEET:

Title:

Analisi agli elementi finiti per il controllo della fessurazione nei muri in c.a. soggetti a fenomeni di ritiro (Translated: Finite elements analysis for cracking control of reinforced concrete walls prone to shrinkage)

Authors:

A.   Meda, G. Plizzari, C. Zanotti, S. Cangiano

Prepared for:

University of Rome, University of Brescia, Roma, Italcementi Group

Publication type:

Conference proceedings

Date of publication:

June 2009

Publisher:

IGF XX National Conference

ISBN:

978-88-95940-25-0

Content summary:

Numerical evaluation on the cracking phenomenon of shrinkable concrete walls

Text length and type:

11 folders containing pictures and graphs

URL for the full text (Italian):

https://books.google.it/books?id=dSaFc_Qmt4sC&pg=PA237&lpg=PA237&dq=Analisi+agli+elementi+finiti+per+il+controllo+della+fessurazione+

*nei+muri+in+c.a.+soggetti+a+fenomeni+di+ritiro&source=bl&ots=q8anW
Q9nlZ&sig=ACfU3U3Evl4d1Gqp_rOVXLT7S69Ag51Esg&hl=it&sa=X&ved=
2ahUKEwiY2IrtwvHnAhVnyoKHXvjAKUQ6AEwAXoECAoQAQ#v=onepage
&q=Analisi%20agli%20elementi%20finiti%20per%20il%20controllo%20d
ella%20fessurazione%20nei%20muri%20in%20c.a.%20soggetti%20a%20
fenomeni%20di%20ritiro&f=false*

3.1.8) Experimental investigation of Insulated Concrete Form (ICF) wall panels under quasi static cyclic load

This study analyzes the ICF formworks behavior under quasi static cyclic load. 60mm thick concrete core and 100mm thick EPS sheets with two different densities (4-8daN/mc) wered used. Once put into the adequate presses, the sample formworks overall measured 1000x500mm.

Firstly, the polystyrene and concrete properties are defined well as the testing and sample curing procedures.

Secondly, the behavior of each wall is described through a series of force-displacement graphs illustrating the hysteresis loop (18 load cycles per sample).

Lastly, the study draws the conclusion that the tested ICF walls with the aforementioned characteristics resist lateral loads quantified in terms of force and displacement.

[Ed. The capacity load framing seems to be rather low, which is presumably due to the relatively thin layer of the concrete wall and to the apparent lack of rebars].

SUMMARY SHEET:

Title:

    Experimental investigation of insulated concrete form (ICF) wall panels under quasi static cyclic load

Authors:

    Arun Solomon, Hema Latha

Prepared for:

    Karunya University, India

Publication type:

    Scientific paper

Date of publication:

January 2018

Publisher:

International Journal of Engineering and Advanced Technology

ISSN:

2249 – 8958

Content summary:

Experimental analysis of ICF walls behavior under cyclic lateral loads

Text length and type:

6 folders containing pictures and graphs

URL for the full text:

https://www.researchgate.net/publication/331638575_Experimental_investigation_of_insulated_concrete_form_ICF_wall_panels_under_quasi_static_cyclic_load

3.1.9) In-plane shear resistance of Insulated Concrete Form walls

This study explains with clarity and detail the concept of shear walls and highlights the advantages they bring to buildings subjected to lateral wind and seismic loads. A distinction is made between light-weighted shear walls (with wood or steel framing) and "heavy" shear walls such as ICFs.

Firstly, the article reports the results of a research program undertaken in the NAHB Research Center. The experimental program is aimed at understanding the effects of openings on ICF shear walls and the influence of the height-to-length aspect ratio on the wall lateral resistance.

To determine such factors, nine ICF shear walls of three different typologies (flat, waffle-grid and screen-grid) were developed. Testing on their structural performance was carried out. The shape, size and rebars of each wall typology are properly illustrated in Table 1. The single tests and wall failure - clearly visible after the insulation removal - are carefully described. Further tables are provided that list the loading and deflection values for each sample.

Lastly, the study proposes the hypothesis of an empirical method (system of equations) relating the opening area ratio to the applied shear force, which varies with the ICF typology. However, the Authors point out that this method may not apply to buildings with very large openings that compromise the shear walls stiffness.

SUMMARY SHEET:

Title:

In-plane shear resistance of Insulated Concrete Form walls

Authors:

NAHB Research Center, Inc. Upper Marlboro, Maryland

Prepared for:

U.S. Department of Housing and Urban Development

Publication type:

Research contribution

Date of publication:

April 2001

Publisher:

PATH

ISSN:

Content summary:

Shear resistance of different ICF walls systems

Text length and type:

50 folders containing pictures and graphs

URL for full text:

https://www.huduser.gov/portal//Publications/PDF/inplane.pdf

3.1.10) Characteristics of expanded polystyrene and its impact on mechanical and thermal performance of Insulated Concrete Form (ICF) system

The study focuses on the impact polystyrene has on the mechanical behavior of an ICF wall, and specifically how the holding effect of polystyrene can change the concrete wall from brittle to ductile.

In the first part of the experimentation, the mechanical behavior of simple EPS specimens undergoes axial compressive strength and flexural tests. The EPS density ranges from 4 to 12daN/mc.

In the second part of the experimentation, six specimen groups of ICFs were examined. ICF samples differ by formwork thickness and polystyrene density. The concrete was identical and measured 60mm. The ICF formworks specimen overall measured 150*200mm.

Further testing was carried out on 500x500mm ICFs in order to determine the R-value (thermal insulation) through a thermal test chamber also accurately described in the article.

As a conclusion, the study presents a series of tables and graphs indicating the ductile behavior of the ICF formworks undergoing mechanical testing (ductility increases with thicker EPS sheets) and the temperature trend history of the tested ICF formworks.

SUMMARY SHEET:

Title:

Characteristics of expanded polystyrene and its impact on mechanical and thermal performance of Insulated Concrete Form (ICF) system

Authors:

Arun Solomon, Hema Latha

Prepared for:

Karunya University, India

Publication type:

Scientific paper

Date of publication:

September 2019

Publisher:

Elsevier ltd

ISSN:

Content summary:

Experimental analysis of the impact of expanded polystyrene properties on mechanical and thermal performance of ICF walls

Text length and type:

10 folders containing pictures and graphs

URL for the full text:

https://www.sciencedirect.com/science/article/pii/S2352012419301900

98

3.1.11) Seismic-proof buildings in developing countries

This particularly interesting article starts by analyzing the safety and antiseismic development perspectives in low-income countries. According to the World Bank, developing countries have experienced 53% of the overall world disasters but have accounted for 93% of the casualties.

Firstly, the study states that by 2050, a billion of new house units is expected to be required to face the ever-increasing world population. Therefore, the building industry needs to take on the challenge of making the construction process safer in full compliance with the current regulations. It must also be taken into account that most low-income countries have adopted regulations borrowed from other countries. This approach becomes a limitation since mature regulamentary systems cannot be integrated in developing countries, that for their part need a specific set of adequate rules. Once these rules have been adapted to the local culture, economy, society and institutions, the compliance may be affected.

It appeared that modern reinforced concrete structures in developing countries did not resist the recent seismic hazards. A few cases of relevant strategic building collapses were reported. This is due to the fact that integral safety is guaranteed by the building ductile behavior given by specific details.

Firstly, the history of frame structures is retraced. The article highlights their insufficient inherent resistance properties and the high ductility required to guarantee an adequate seismic response.

Secondly, new international seismic-proof design philosophies are defined. The article explains that in areas prone to frequent seismic hazards, the main goal should be that to create buildings that remain fully operational (Fully Operational performance level).

Additionally, an in-depth analysis of "lateral systems" for tall buildings is offered. Seismic-proof buildings are distinguished between "shear walls" and rigid concrete walls. Moreover, the so-called framed tube structures are analyzed. These structures allow to create extremely high buildings with high seismic safety standards.

Lastly, a similarity between reinforced concrete wall systems and PierLuigi Nervi's ferrocement is underlined.

ICFs are taken as an example of great seismic-proof reinforced concrete walls and their properties are carefully described shortly thereafter.

SUMMARY SHEET:

Title:

Seismic-proof buildings in developing countries

Authors:

Vittoria Laghi, Michele Palermo, Tomaso Trombetti, et al.

Prepared for:

Department of Civil and Environmental Engineering, University of Bologna

Publication type:

Scientific paper

Date of publication:

August 2017

Publisher:

Frontiers

ISSN:

Content summary:

Seismic-proof systems history and analysis for the construction of tall buildings in developing countries

Text length and type:

12 folders containing graphs and pictures

URL for full text:

https://www.frontiersin.org/articles/10.3389/fbuil.2017.00049/full

3.1.12) Further references

The present paragraph is aimed at listing further studies on reinforced concrete walls conducted by international researchers. These works have not been analyzed in the present book due both to lack of space and their indirect relation to flat ICF walls systems, on which this work is explicitly focused.
Here follows a list indicating the title, author's name, institution and date of publication of further studies the reader may be interested in:

- Experimental investigation of lateral cyclic behavior of wood-based screen-grid Insulated Concrete Form walls, John Stuart Garth, Portland State University, 2014
- Measurement systems for determining formwork pressure of highly-flowable concrete, Kamal H. Khayat Æ Joseph J. Assaad, Universite' de Sherbrooke, Sherbrooke, QC, Canada, 2006
- Response modification factor due to ductility of screen-grid ICF wall system in high seismic risk zones, Pouria Asadi, Seyed Mehdi Zahrai, University of Rhode Island, 2016
- Mechanical model for seismic response assessment of lightly reinforced concrete walls, Alberto Pavese et al., University of Pavia, 2016
- Portable pressure device to evaluate lateral formwork pressure exerted by fresh concrete, Ahmed F. Omran et al., Département de Génie Civil, Université de Sherbrooke, 2500 Blvd. de l'Université, Sherbrooke, QC, Canada, 2013
- Cyclic behavior of screen grid Insulated Concrete Form components, Carl Scott Werner, Portland State University, 2010
- Design of a shaking table test on a 3-storey building composed of cast-in-situ concrete walls, Tomaso Trombetti et al., University of Bologna, 2012

- *In-plane shear behaviour of thin low reinforced concrete panels for earthquake re-construction,* Tomaso Trombetti et al., University of Bologna, 2011

- **Pressure of concrete on formwork,** Andreas Leemann, Empa - Swiss Federal Laboratories for Materials Science and Technology, 2006

- **Preliminary interpretation of shaking-table response of a full-scale 3-storey building composed of thin reinforced concrete sandwich walls,** Tomaso Trombetti et al., University of Bologna, 2014

- **Aftershock response of RC buildings in Santiago, Chile, succeeding the magnitude 8.8 Maule earthquake,** Anne Lemnitzer, University of CA, Irvine, Dept. of Civil & Environmental Engineering,2013

- **Collapse Assessment of the Alto Rio Building in the 2010 Chile Earthquake,** Zeynep T. Değer, Istanbul Technical University, 2014

3.2) Energetic analyses

Along with the structural characteristics, the energetic properties of ICFs are the strong point of this system. Therefore, a number of researchers focused on investigating the buildings thermal performance and comparing it to that of traditional construction systems. Interestingly enough, these examinations were undertaken in different geographic areas around the world under very different climates. However, ICFs were found to guarantee a high thermohygrometric performance in all the given conditions, and remarkable energy saving was observed during both summer and winter. Wood-framed systems appear to be astonishingly more energy-intensive than ICFs.

Here follows a list of the selected works on the topic:

3.2.1) How Insulating Concrete Form vs conventional construction of exterior walls affects whole building energy consumption: results from a field study and simulation of side-by-side houses

3.2.2) Thermal analysis of Insulated Concrete Form (ICF) walls

3.2.3) Assessment of ICF energy saving potential in whole building performance simulation tools

3.2.4) Feasibility of using Insulated Concrete Forms in hot and humid climate

3.2.5) Energy use in residential housing, a comparison of Insulating Concrete Form and wood frame walls

3.2.6) Monitored thermal performance of ICF walls in MURBs

3.2.7) Measured cooling performance of two-story homes in Dallas, Texas; Insulated Concrete Form versus frame construction

3.2.8) Further references

3.2.1) How Insulating Concrete Form vs. conventional construction of exterior walls affects whole building energy consumption: results from a field study and simulation of side-by-side houses

This article examines the energy consumption of two side-by-side 100-square-meter houses - namely an ICF building and a wood-frame one. The houses are located in the U.S., in East Tennessee.

Firstly, the weather was monitored over 11 months from 2000 to 2001. For the twelfth month, further weather data was collected analytically in order to validate the full annual energy usage.

Secondly, the study provides results on the ICF system energy saving. The unoccupied ICF building saved 7.5% more energy than the conventional houses, whereas when occupied, the ICF building saved 9.2% more energy.

Furthermore, the article highlights that wider temperature fluctuations were observed inside the traditional building through the seasons, especially when no further energy supply was provided. This shows that thermally massive buildings like ICFs bring further advantages beyond energy saving over conventional construction methods.

Lastly, the air tightness of each building underwent specific in-splace testing. As the results show, ICFs exhibit higher values by 10%.

It is worth noting that the transparency over data provided makes this study analytic and impartial. Detailed tables indicating full thermal data on the examined building materials is shown.

Also, the article offers an interesting validated predictive model showing the energy savings of these constructive systems (ICF and conventional) with weather data from various climatic areas all over the U.S.. These results are shown in a table indicating that the overall energy savings guaranteed by ICFs range from 5.5% (Miami) to 8.5% (Phoenix).

SUMMARY SHEET:

Title:

How Insulating Concrete Form vs. conventional construction of exterior walls affects whole building energy consumption: results from a field study and simulation of side-by-side houses

Authors:

Thomas W. Petrie, Jan Kosny, André O. Desjarlais, Jerald A. Atchley, Phillip W. Childs, Mark P. Ternes and Jeffrey E. Christian

Prepared for:

Oak Ridge National Laboratory

Publication type:

Scientific paper

Date of publication:

Publisher:

ISSN:

Content summary:

Comparative analysis of the energy savings of wood-frame vs traditional buildings in different climatic conditions around the U.S.

Text length and type:

12 folders containing graphs

URL for the full text:

https://www.aceee.org/files/proceedings/2002/data/papers/SS02_Panel1_Paper19.pdf

3.2.2) Thermal analysis of Insulated Concrete Form (ICF) walls

This study deals with the walls performance as thermal energy storage. More precisely, the article analyzes the case study of an innovative building design in which water pipes were embedded inside the reinforced concrete of an ICF wall to detect its fluid/concrete heat transfer. The insulation of the ICF walls was found to be ideal to keep the heat inside, and therefore to maintain the fluid temperature. At this point, the thermal energy stored inside the walls can be used to heat the house floor, this way enhancing the house thermal performance. As a result, besides performing a structural and enveloping function, walls become a thermal flywheel with no need for further water tanks, this way optimizing the wall usage. In order to define the adequate amount of concrete needed to produce the thermal flywheel effect, the researchers simulated a numerical model (concrete wall thickness of 6 inches, concrete density of 2000daN/mc, specific heat capacity of 1000J/kg*K, and thermal conductivity of 1.13W/m*K) with ½ inch pipes and different temperatures. This model allowed to properly define the overall heat transfer process.

SUMMARY SHEET:

Title:

    Thermal analysis of Insulated Concrete Form (ICF) walls

Authors:

    Navid Ekrami et al.

Prepared for:

    Ryerson University, 350 Victoria Street, Toronto, Ontario, Canada

Publication type:

    Scientific paper

Date of publication:

2015

Publisher:

Elsevier Ltd.

ISSN:

Content summary:

Thermal analysis of ICF as a thermal flywheel

Text length and type:

7 folders containing graphs

URL for the full text:

https://www.researchgate.net/publication/282417355_Thermal_Analysis_of_Insulated_Concrete_Form_ICF_Walls

3.2.3) Assessment of ICF energy saving potential in whole building performance simulation tools

Over the past years, little change was introduced to constructive technologies in the UK. However, a few recent surveys highlighted the ever-increasing awareness towards the advantages of innovative energy-saving sustainable technologies. In this framework, this paper aims to analyze the potential of the ICF constructive system in the UK.

In the beginning, a careful examination of the main benefits is offered. As far as thermal properties are concerned, the Authors make reference to data coming from the American publications in the field. Many comparative analyses have been provided by American researchers in which ICF buildings set in different representative locations are compared to conventional framing. These studies show that for every location examined, ICFs guarantee higher thermal insulation and air tightness. Even the Life Cycle Assessment (LCA) found that despite the slightly higher environmental footprint made during the formwork manufacture phase, ICF systems have a low environmental impact.

Similar results were provided by studies on ICFs thermal mass conducted in Canada as well.

The goal of this article is to provide numerical simulations of the ICF system energy consumption compared to that of a high thermal mass system and a low thermal mass system under the same environmental conditions.

The advantages of ICFs evidenced by these simulation results are shown in a series of tables and graphs.

SUMMARY SHEET:

Title:

Assessment of ICF energy saving potential in whole building performance simulation tools

Authors:

Eirini Mantesi et al.

Prepared for:

School of Civil and Building Engineering, Loughborough University, Leicestershire

Publication type:

Conference paper

Date of publication:

2015

Publisher:

ISSN:

Content summary:

ICF systems energy savings

Text length and type:

9 folders containing pictures and graphs

URL for the full text:

https://repository.lboro.ac.uk/articles/Assessment_of_ICF_energy_saving_potential_in_whole_building_performance_simulation_tools/9431732

3.2.4) Feasibility of using Insulated Concrete Forms in hot and humid climate

This paper compares the walls performance of an ICF and a conventional building both located in the Sultanate of Oman - where energy is particularly cost-effective. The climate of Oman is peculiarly characterized by very moist summers (80% humidity) that disrupt the regular course of social activities, and during which the air conditioning loads lead to electric energy absorption peaks.

Firstly, the paper goes through previous comparative analyses undertaken all over the world. From these works, the energy savings guaranteed by ICFs exceed those of than conventional buildings by 20% (such as in the U.S.).

Secondly, the paper analyzes two houses built on the Omani soil, composed of concrete blocks (thermal conductivity of 1.8W/m*K) and ICF formworks ((0.40W/m*K) respectively. A number of tables and graphs indicate that the ICF system energy savings were up to 40% more than those of the traditional building.

Lastly, the paper draws the conclusion that the energy demand for air conditioning in an ICF building is lower than in a traditional building by 36%. Therefore, ICFs guarantee such a lower power purchase as to offset the superior construction costs of ICFs in this specific area.

After assessing the feasibility of ICFs in hot and humid climates, however, the Authors raise a few doubts over the behavior of concrete along coastal areas, whose durability should therefore undergo further testing.

SUMMARY SHEET:

Title:

   Feasibility of using Insulated Concrete Forms in hot and humid climate

Authors:

   K.P. Ramachandran et al.

Prepared for:

Caledonian College of Engineering, Muscat, Sultanate of Oman

Publication type:

Scientific paper

Date of publication:

2014

Publisher:

International Journal of civil engineering and technology

ISSN:

0976 – 6308

Content summary:

Energy savings of ICFs in humid climates

Text length and type:

9 folders

URL for the full text:

https://www.semanticscholar.org/paper/FEASIBILITY-OF-USING-INSULATED-CONCRETE-FORMS-IN-Ramachandran/f74426a3bfb2c60db22757ba076abbfb8d0815b4

3.2.5) Energy use in residential housing, a comparison of Insulating Concrete Form and wood frame walls

This article analyzes the energetic performance of a building modeled through the numeric simulations of a calculation engine. This residential building is a detached two-story 228-square-meter house with a contemporary design, located around the U.S. in five different areas representing the various climates, namely:

- Phoenix, Arizona (hot dry climate);
- Miami, Florida (hot humid climate);
- Seattle and Washington (moderate climates);
- Chicago, Illinois (cold climate).

For each location, three building variations were modeled: wood-frame walls, ICF walls, and code-matching non-mass walls complying with the minimum energy local code requirements. Other building elements (windows, installations, partition walls, and so on) were all alike.

As the study shows, the average energy consumption reduction of ALL the ICF walls ranged from 8% to 19% if compared to code-matching walls, whereas the average energy consumption was from 5 to 9% below than that of wood-frame walls. As far as the overall system capacity defined by the software was concerned, the energy saving gap in favour of ICFs was shown to be even larger.

Additionally, the paper reports weather data from the different climatic areas and further specifics on the characteristics of each building as well as comparative graphs for each constructive solution.

SUMMARY SHEET:

Title:

Energy use in residential housing, a comparison of Insulating Concrete Form and wood frame walls

Authors:

John Gajda and Martha VanGeem

Prepared for:

Construction Technology Laboratories Inc. (CTL) Orchard Road, Skokie, U.S.A.

Publication type:

Scientific paper

Date of publication:

Publisher:

Portland Cement Association

ISSN:

Content summary:

Energy savings of an ICF building vs other constructive systems under different climates

Text length and type:

17 folders

URL for the full text:

https://www.cement.org/docs/default-source/sustainabilty2/energy-use-in-residential-housing---a-comparison-of-insulating-concrete-form-and-wood-frame-walls.pdf?sfvrsn=4

3.2.6) Monitored thermal performance of ICF walls in MURBs

This article is a funded piece of research aiming to measure the thermal performance and air tightness of an ICF housing complex, and assess the implications of thermal performance on the installments sizing. In order to do so, the thermal performance of the walls was monitored for a full year. Monitoring was carried out through specific sensors installed onto the interior and exterior surface of polystyrene as well as on the interface between both the inner and the outer layers of the insulation and the concrete. In addition to sensors, further thermographic scanning was adopted to detect any thermal bridges.

The article accurately lists the structural and thermal properties of the materials employed for the walls. Furthermore, the numerical model used for the analyses is fully illustrated. Also, this model analyzes the temperature of constructive intersections.

The study allowed to verify the perfect air tightness of the building, the negligible nature of its thermal bridges and more generally, the excellent energetic properties of the house to the advantage of the occupants.

All simulations were carried out by a specific software which allowed to calculate the yearly energetic costs for each model, from electricity bills to natural gas consumption.

SUMMARY SHEET:

Title:

Monitored Thermal Performance of ICF walls in MURBs

Authors:

Duncan Hill

Prepared for:

CMHC (Canada Mortgage and Housing Corporation)

Publication type:

Research contribution

Date of publication:

2007

Publisher:

CMHC

ISSN:

Content summary:

Monitoring of the actual thermal performance of an ICF building

Text length and type:

6 folders containing pictures and graphs

URL for the full text:

https://docplayer.net/34088661-Monitored-thermal-performance-of-icf-walls-in-murbs.html

3.2.7) Measured cooling performance of two-story homes in Dallas, Texas, Insulated Concrete Form versus frame construction

This study aims to monitor the thermal behavior of four two-story buildings during summertime in Dallas. Energy consumption for air conditioning was measured for each building - namely two wood-frame and two ICFs. The buildings were similar yet not identical, especially in their use. Therefore, the article adequately adjusted the results according to these differences. The different color of the exterior brick veneer was one of the reported differences, since it was found to influence the solar energy absorption level (one of the ICFs was of lighter color).

By analyzing this data, the Authors drew the conclusion that ICFs can reduce energy demand up to 19% over traditional buildings. These results are fully in line with those of other international researchers.

The study provides pictures and graphs on each building typology as well as the walls stratigraphy (all buildings have exterior brick veneer). However, the Authors also point out that further comparative studies on buildings fully identical in construction (except for the walls) and use are needed.

SUMMARY SHEET:

Title:

Measured cooling performance of two-story homes in Dallas, Texas, Insulated Concrete Form versus frame construction

Authors:

Dave Chasar et al.

Prepared for:

Florida Solar Energy Center

Publication type:

Research contribution

Date of publication:

Publisher:

ISSN:

Content summary:

Energy consumption for ICF vs wood-frame buildings cooling

Text length and type:

10 folders

URL for the full text:

https://www.researchgate.net/publication/26901094_Measured_Cooling_Performance_of_Two-story_Homes_in_Dallas_Texas_Insulated_Concrete_Form_Versus_Frame_Construction

3.2.8) Further references

The space and deadlines of the present book make it impossible to undertake careful analysis of every study found on the Internet. As in Paragraph 3.1, it was therefore decided to list the titles of those works considered to be more specialized and therefore more difficult to summarize.
Readers interested in the topic can freely proceed to read these works in autonomy.

- ***Applications of active hollow core slabs and insulated concrete foam walls as thermal storage in cold climate residential building,*** *Navid Ekrami et al., Ryerson University, Canada, 2015*
- ***Field energy performance of an Insulating Concrete Form (ICF) wall,*** *Maref W. et al., National Research Council Canada, 2012*
- ***Benchmarking 3D thermal model against field measurement on the thermal response of an Insulating Concrete Form (ICF) wall in cold climate,*** *Saber, H.et al., National Research Council, Canada, 2010*
- ***Hygrothermal performance assessment of ICF walls with different moisture control strategies and wall designs,*** *Emishaw Iffa et al., British Columbia Institute of Technology, Canada*
- ***Air leakage of Insulated Concrete Form houses,*** *Hannah Durschlag, Massachusetts Institute of Technology, 2012*
- ***Insulated Concrete Form walls integrated with mechanical systems in a cold climate test house,***
 D. Mallay et al., 2014

3.3) Constructive analyses

The present paragraph goes through a series of international studies dealing with the constructive problems that may arise using the ICF system. These scientific works mainly focus on the casting process and the evaluation of concrete integrity. Throughout these studies, a few operative procedures of ICF walls are outlined as well as several investigation and reconstruction technologies. These are the only studies found on the topic. It appears from the articles that most constructive problems are mainly due to on-site blunders (reinforcement placing, concrete pouring and consolidation) that are easily prevented by following the project prescriptions and operational guidelines on ICFs. These guidelines are partially reported in Chapter 1 of the present volume as well as in the specialized works in Chapter 2. Here follows a list of the selected scientific studies concerning ICFs construction:

> *3.3.1) Concrete consolidation and the potential for voids in ICF walls*
>
> *3.3.2) A case study on construction defects of reinforced concrete walls with Insulated Concrete Forms*
>
> *3.3.3) Inspection of Insulated Concrete Form walls with ground penetrating radar*
>
> *3.3.4) Job-built Insulated Concrete Forms (ICF) for building construction*

3.3.1) Concrete consolidation and the potential for voids in ICF walls

This interesting report mainly focuses on the constructive defects of ICF walls caused by different concrete casting processes and vibration.

The article starts by examining a high number of ICF wall samples differing in shape and formwork type (flat-panel, screen-grid or waffle-grid), concrete mix designs (different slumps, self-consolidating concrete), and vibration mode (internal vibration, external vibration).

Firstly, the insulation was removed on one side to allow the casting verification and highlight any constructive defects. Voids were documented especially in lintels and corners, where rebar congestion can lead to significant voiding.

Secondly, a nondestructive testing technique using an impulse radar to verify the proper ICF wall casting process is described.

According to the study results, internal vibration was found to be the most reliable method to guarantee the adequate ICF formwork casting and concrete segregation. However, internal vibration needs to be distributed onto the wall surface at regular intervals in order for the zones of influence not to overlap. As an alternative, optimal concrete consistency was guaranteed by the self-consolidating concrete mix. However, close attention must be paid to the concrete pressure on the formworks, which may lead to the blowout of the formworks themselves.

Lastly, environmental factors such as temperature or water addition were indicated as a negative influence on the casted concrete, since they help create voids and reduce the concrete strength.

The article is full of technical drawings of the investigated walls showing the distribution of the potential voids left during casting.

SUMMARY SHEET:

Title:

Concrete consolidation and the potential for voids in ICF walls

Authors:

John Gajda and Amy M. Dowell

Prepared for:

Construction Technology Laboratories, Inc., 5400 Old Orchard Road, Skokie

Publication type:

Research contribution

Date of publication:

2003

Publisher:

Portland Cement Association

ISBN:

0-89312-232-7

Content summary:

Mistakes in the ICF wall in-place pouring process

Text length and type:

16 folders containing pictures

URL for the full text:

various

122

3.3.2) A case study on construction defects of reinforced concrete walls with Insulated Concrete Forms

This articles analyzes the case study of an ICF building whose rebars appeared to have been inadequately placed. The building structure also presented segregation and voids.

This led to the order by the engineer in charge to remove large portions of insulation to uncover the concrete core and repair the voids through carbon fiber installations. In selected locations, a ground penetrating radar was also used. Both operations were time-consuming and cost-intensive since they covered large portions of the walls. However, they successfully guaranteed the full building recovery and operability without structural demolition.

The defects were traced back to gross executive errors such as incorrect aggregate size, incorrect or lacking vibration practices, and insufficient spacing between the reinforcing bars.

The study offers simple yet important recommendations to adopt before casting the ICf walls - since the defects detection gets much more difficult once the formworks have been cast.

The building was a federal property, and therefore was not subject to the usual codes for residential construction (which means it did not undergo the usual executive verifications during the construction process). This might be one of the reasons why the critical issues were not detected earlier. Since the building needed to comply with a few stringent requirements including blast resistance, it was then necessary to systematically recover the building according to the previously stated modes.

The case study offers plenty of pictures of inadequate wall casting.

SUMMARY SHEET:

Title:

A case study on construction defects of reinforced concrete walls with Insulated Concrete Forms

Authors:

David B. Peraza

Prepared for:

M.ASCE

Publication type:

Conference paper

Date of publication:

2015

Publisher:

ISBN:

Content summary:

Defects in ICF reinforced concrete walls

Text length and type:

12 folders with pictures

URL for the full text:

https://www.researchgate.net/publication/301455624_A_Case_Study_on_the_Construction_Defects_of_Reinforced_Concrete_Walls_with_Insulated_Concrete_Forms

3.3.3) Inspection of Insulated Concrete Form walls with ground penetrating radar

Potential defects occurring inside the walls such as honeycombing are one of the constructive aspects requiring the closest attention among ICF structures. Such defects are usually difficult to detect due to the external insulation that serves as a permanent envelope for the wall. The study presents a radar technique called Ground Penetrating Radar (GPR). The GPR allows to investigate the reinforced concrete walls consistency through acoustic waves whose reflections increase if the surface they meet is flat and solid. This method only allows to detect significant defects whose diameter exceeds 80-100mm. However, it should be noted that defects with inferior diameter do not actually affect the wall performance.

Three ICF wall specimens were developed. The walls were adequately reinforced. In order to simulate constructive defects, artificial voids were created by means of pipes pre-inserted during casting and then removed.

The results of the GPR inspection are shown through radargrams. These radargram assess the reliability of the provided data, which show the usability of this defect-detecting technology.

In conclusion, the GPR was found to be a reliable means to inspect ICF walls in spite of their insulation, reinforcement, and plastic spacers, that are likely to interfere with the waves transmission. Trained operators are therefore required for this specific inspection.

The study offers plenty of picture and figures illustrating this inspection method and the return data mode.

SUMMARY SHEET:

Title:

Inspection of Insulated Concrete Form walls with ground penetrating radar

Authors:

Cherif Amer-Yahia, Todd Majidzadeh

Prepared for:

Resource International, Inc., 6350 Presidential Gateway, Columbus, OH 43231, United States

Publication type:

Scientific paper

Date of publication:

2010

Publisher:

Construction and Building Materials

ISBN:

Content summary:

Radar inspection of ICF walls

Text length and type:

13 folders containing pictures

URL for the full text:

https://www.researchgate.net/publication/257307121_Inspection_of_Insulated_Concrete_Form_walls_with_Ground_Penetrating_Radar

3.3.4) Job-built Insulated Concrete Forms (ICF) for building construction

This study starts from the assumption that ICFs are a promising and appealing system due to the advantages they have over traditional building. According to the National Association of Home Builders (NAHB), ICFs cover 3.0% of the overall building marketplace in the U.S. (NAHB, 2005). Over the recent years, this percentage saw a major increase. As the study points out, however, the initial costs of ICFs (ed. in the American marketplace) exceed those of traditional wood-framing (ed. with lower performance) by 3-5%. Therefore, a new formwork typology is introduced in which the EPS sheets are interconnected by threaded glass fiber reinforced polymers spacers. Since these spacers show improved mechanical properties, their size is slightly inferior than usual and they therefore have a lower impact on the concrete vibration and consolidation processes.

In order to evaluate this new system, and specifically its construction time and structural performance, a full-scale wall sample with self-consolidating concrete was developed. According to the Authors, self-consolidating concrete usage was allowed namely by the reinforced spacers, which held the liquid concrete pressure without any risks of breakage.

Thanks to the spacer properties and resistance *[Ed. properly shown through tables]*, safer pouring in place operations at faster rate were made possible, and therefore construction costs reduction was guaranteed.

SUMMARY SHEET:

Title:

Job-built Insulated Concrete Forms (ICF) for Building construction

Authors:

Afshin Hatami and George Morcous

Prepared for:

University of Nebraska Lincoln

Publication type:

Conference proceedings

Date of publication:

2011

Publisher:

Associated Schools of Construction

ISBN:

Content summary:

Introduction of reinforced spacers in the ICF system

Text length and type:

8 folders containing graphs

URL for the full text:

http://ascpro0.ascweb.org/archives/cd/2011/paper/CPRT343002011.pdf

3.4) Environmental sustainability analyses

Nowadays, the need for construction projects with a low social and environmental impact is a much-debated topic due both to the freshly introduced regulations on sustainability and the international community's increasing awareness towards resource depletion.

This awareness has encouraged the so-called "green building", whose main goals are energy efficiency, health improvement, occupants' comfort, and materials recyclability.

As it is clear, green building is only made possible by the proper constructive method and fundamental design factors such as orientation, daylighting, natural ventilation, and orientation, soleggiamento, natural ventilation, and an adequate plant engineering.

The ICF construction system fully complies with all of these requirements, as evidenced by the following studies:

3.4.1) Recycled foam and cement composites in Insulating Concrete Forms

3.4.2) An analysis of Insulated Concrete Forms for use in sustainable military construction

3.4.1) Recycled foam and cement composites in Insulating Concrete Forms

This article starts by stressing the substantial changes the building industry is undergoing due to the increasing awareness towards resource depletion.

ICF systems already are a great example of sustainable building thanks to the construction process optimization they guarantee as well as their reduced energy consumption. However, further improvements can be introduced to this system to make it even more sustainable. For example, partially recycled foam and recycled concrete could be increasingly employed.

As shown in the article, out of the 40-50 ICF manufacturers operating all over North America, only 3 have recently substituted formworks with recycled materials for those with virgin polystyrene.

The article provides a table listing these three recycled formwork manufacturers. This data shows that only one of them actually produced recycled polystyrene formworks (namely, Rastra), since the others produce timber-concrete forms.

Lastly, the article closely examines the properties of the materials composing the Rastra formworks. As a conclusion, the Authors highlight that the number of manufacturers for the building industry that are taking ever greater interest in recycled materials is increasing.

SUMMARY SHEET:

Title:

  Recycled foam and cement composites in Insulating Concrete Forms

Authors:

  Richard Boser, Mr. Tory Ragsdale, and Dr. Charles Duvel

Prepared for:

  Illinois State University

Publication type:

Scientific paper

Date of publication:

2002

Publisher:

Journal of Industrial Technology

ISBN:

Content summary:

Recycled materials for ICF walls

Text length and type:

5 folders containing pictures

URL for the full text:

https://www.researchgate.net/publication/294513156_Recycled_foam_and_cement_composites_in_insulating_concrete_forms

3.4.2) An analysis of Insulated Concrete Forms for use in sustainable military construction

This research arises from the need for higher energy-efficient constructive technologies and sustainable materials expressed by the American Department of Defense. Therefore, the study closely examines the ICF constructive system and its advantages. Although this system tends to cost slightly more in the U.S., it indeed offers great potential for energy consumption reduction.

Through an energy modeling software, numerical simulations of the energy savings for different military buildings located in the U.S. were carried out. Besides the energetic analysis, a few military buildings also underwent the Life Cycle Cost Analysis. This led to the conclusion that ICFs deserve to be considered one of the most energy-efficient constructive systems for military construction.

Lastly, the Author stresses the fact that the contractors' lack of knowledge of this system results in a barrier for the system implementation.

SUMMARY SHEET:

Title:

 An analysis of Insulated Concrete Forms for use in sustainable military construction

Author:

 Rebecca L. Ponder

Prepared for:

 Department of the air force air University "AIR FORCE INSTITUTE OF TECHNOLOGY", Ohio, USA

Publication type:

 Thesis

Date of publication:

2014

Publisher:

AIR FORCE INSTITUTE OF TECHNOLOGY

ISBN:

Content summary:

Sustainable use of ICFs in military implementations

Text length and type:

148 folders containing pictures and graphs

URL for the full text:

https://www.semanticscholar.org/paper/An-Analysis-of-Insulated-Concrete-Forms-for-use-in-Ponder/a380de01f2bcf7e537a7d105c642b105ed183760

3.5) Exceptional loading analyses

In Europe, building design is profoundly different from the American one. This is mainly due to three factors. Firstly, different regulations apply in America. Secondly, the American building industry is characterized by a different design approach, which is prescriptive and oversimplifying as far as simple constructions are concerned. But most significantly, buildings in the U.S. have to resist different natural hazards than in Europe. Specifically, buildings need to resist tornadoes, which threaten the American coastal areas more frequently than earthquakes, if not more severely.

As accurately reported by the following works, ICF structures can successfully resist the effects produced by tornadoes - that is to say, not merely the forces they exert, but also indirect effects such as wind projectiles travelling at a high rate of speed. The following studies also deal with further exceptional loads as blasts. Thanks to their external insulation, ICFs were found to fully resist the shockwave generated from these eruptions as well.

Here follows a list of the works reviewed on the topic:

3.5.1) ICF as blast-resistant barriers

3.5.2) Insulating Concrete Forms as a blast resistant building material

3.5.3) Investigation of wind projectile resistance of Insulating Concrete Form homes

3.5.1) ICF as blast-resistant barriers

This study focuses on the ability of EPS formworks to absorb the compression shock wave generated from explosive blasts.

Moreover, the study highlights the restraining function of the plastic spacers that interconnect the insulating layers. After the blast, the spacers prevent the walls from potential cracking, thus stopping any concrete fragments from piercing the building interior walls.

The study mentions a series of tests dating back to 2003 in which specific amounts of dynamite were used to blow up cubic-shaped ICF walls. The wall samples were put with different angles at different distances from 3 blast points located against an infinitely rigid steel barrier. Afterwards, the walls underwent close examination to analyze the blast effects.

Through this approach, it was possible to assess that the outer EPS layer "protects" the structural part by absorbing the whockwave and therefore reducing the destructive effect.

The numerical model derived by the experimental results is provided.

Lastly, the study suggests that further experimentation would be needed to complete the investigation.

SUMMARY SHEET:

Title:

ICF as blast-resistant barriers

Authors:

R. F. Oleck et al.

Prepared for:

McLaren Engineering Group, USA

Publication type:

Conference proceeding

Date of publication:

2012

Publisher:

ISBN:

Content summary:

Blast-induced effects on ICFs

Text length and type:

11 folders containing graphs

URL for the full text:

https://www.researchgate.net/publication/268589950_Insulated_Concr ete_Forms_ICF_As_Blast-Resistant_Barriers

3.5.2) Insulating Concrete Forms as a blast resistant building material

In this work, a report is given of the same experimental program described in the previous paper. However, this report offers a more detailed summary and plenty of pictures of the testing samples. As described through the study, the tests were carried out in a specific Navy lab on initiative of some category associations (representative for ICFs as well as for concrete). The provided pictures give the best description of the effects produced by the violent blast that hit the ICF samples. The samples were composed of a number of boxes created with the ICF technology (three closed sides plus a closure slab, with a free side at the rear to allow inspection) and put at specific distances from the blast point.

The blast point (to which many explosive charges were applied) was located against an infinitely rigid steel bar whose function was that to amplify the eruption. Once the cycling explosions had taken place, the EPS layers were removed in order to inspect the concrete surface. In every sample, the concrete surface remained almost intact. Even the box set closest to the blast, whose EPS was affected, presented minor damage.

Blast testing is considered to be quite strategic due to the recent terrorist attacks which now impose to consider the possibility of blast damage or at least promote those systems that guarantee resistance to it.

SUMMARY SHEET:

Title:

Insulating Concrete Forms as a blast resistant building material

Authors:

ICFA (Insulating Concrete Forms Association)

Prepared for:

ICFA (Insulating Concrete Forms Association)

Publication type:

Test report

Date of publication:

2003

Publisher:

ISBN:

Content summary:

ICF resistance to explosions

Text length and type;

7 folders containing graphs

URL for the full text:

https://www.researchgate.net/publication/268589950_Insulated_Concr ete_Forms_ICF_As_Blast-Resistant_Barriers

3.5.3) Investigation of wind projectile resistance of Insulating Concrete Form homes

Tornado projectiles are one of the most unpredictable loadings threatening buildings located along the American coastal areas. The ICF system undoubtedly offers higher resistance than traditional wood-frame buildings. Therefore, the ICF wall performance under this specific load was investigated in the Wind Engineering Research Center of the Texas Tech University. The testing was run with ten wall samples among whom ICFs, wood-frame, and reinforced concrete walls. Wooden poles of specific size and weight were thrown onto the walls at high speed rate (119mph) through an air cannon. The testing procedure simulated the load of a tornado whose wind speed amounts to 250mph. This value is the highest peak a tornado can reach in most of America.

The testing showed that unlike wood-frame walls, the ICF and typical reinforced concrete interior walls were in no way damaged. The wooden poles were shattered after penetrating the exterior wall, whereas they easily penetrated the wood and steel-frame walls in spite of their more resistant covering materials.

The paper is full of pictures of the sample walls, which allow to see the real consequences brought by the projectiles on every wall typology.

SUMMARY SHEET:

Title:

Investigation of wind projectile resistance of Insulating Concrete Form homes

Authors:

Ernst W. Kiesling, Ph.D., P.E. and Russell Carter

Prepared for:

Texas Tech University, Department of Civil Engineering

Publication type:

Lab report

Date of publication:

Publisher:

Research and Development Bulletin

ISBN:

Content summary:

Evaluation of ICF wall resistance to projectiles

Text length and type:

23 folders containing graphs

URL for the full text:

http://wijewardane.com/images/PDF/ICF_Benefits.pdf

3.6) Economic analyses

The choice of the constructive system is one of the most delicate moments in the whole building process. This is due to the fact that all the subsequent planning, organization, construction and maintenance phases depend on the system adopted.

However, the system choice is often influenced by the economic factor. As far as innovative technologies like ICFs are concerned, the costs are not even certain. This uncertainty can discourage operators, who may prefer to resort to traditional systems that are nonetheless inferior on many other levels.

It should be noted that construction costs strictly depend on the geographic area, since in different areas, operators incur different labor and material costs. With specific reference to ICFs, it has been found that in Europe (especially in Italy and Mediterranean Europe), costs are lower than those of traditional systems (brick or wood). A comparative costs analysis is available in the volume *Edifici sismoresistenti progettati con le NTC 2018* mentioned in Chapter 2.

In the U.S., as shown in the following studies, the costs involved with ICFs are similar to those of conventional construction methods. However, the construction method conventionally used in the U.S. is wood framing, whose performance is among the lowest a constructive system can have.

Therefore, a series of studies analyzing the ICF system from an economic standpoint is listed. The overall collection offers a global standpoint on the topic - which seems to lead to the conclusion that for the same costs, ICFs provide a wide range of additional advantages over traditional building.

Here follows the selected articles list:

3.6.1) Comparative analysis of houses built from Insulating Concrete Formwork – case study

3.6.2) Costs and benefits of Insulating Concrete Forms for residential construction

3.6.3) Assessment on the usage of Insulated Concrete Form

3.6.1) Comparative analysis of houses built from Insulating Concrete Formwork – case study

This article aims to compare the costs of an ICF to those of a traditional masonry building. Both houses are located in Slovakia, where modern constructive systems with partially prefabricated elements such as ICFs are increasingly spreading.

Firstly, the properties of two specific ICF systems and the Porotherm masonry system are compared. A questionnaire survey was conducted on a sample of people who were asked about the main factors affecting their constructive system choice. The survey highlighted that the main factor influencing people's choice is namely "construction costs" (over 50%).

Secondly, the study analyzes the construction costs of a residential building according three variants - the two ICF systems selected and Porotherm.

According to Table 3 in the article, Porotherm was found to be the most cost-intensive technology both for direct construction and maintenance costs.

SUMMARY SHEET:

Title:

    Comparative analysis of houses built from Insulating Concrete Formwork – case study

Authors:

    Daniela Mačková, Marcela Spišáková

Prepared for:

    Technical University of Košice

Publication type:

    Scientific paper

Date of publication:

2015

Publisher:

JOURNAL OF CIVIL ENGINEERING

ISBN:

Content summary:

Costs analysis of ICF vs masonry buildings

Text length and type:

8 folders containing pictures and graphs

URL for the full text:

https://www.researchgate.net/publication/285674014_Comparative_Analysis_of_Houses_Built_from_Insulating_Concrete_Formwork_-_case_Study

3.6.2) Costs and benefits of Insulating Concrete Forms for residential construction

This study, that has already been reviewed in Chapter 2, is once again mentioned in this book due to the specific costs analysis it offers.

Firstly, the article observes the importance of assessing the "best value" of a house. This concept plays a key role in home purchases indeed, and can depend as much on objective considerations as on intangibles, such as comfort, aesthetics, peace, and so on.

Secondly, the ICFs construction and ownership advantages are listed. It should be noted that in the U.S., ICFs tend to cost more than traditional residential houses by 3-5%. However, American traditional residential houses are wood-framed, and therefore the structural vulnerability factor such as lack of tornado resistance should also be taken into account.

According to the comparative costs analysis provided by the article, a 2,500-square-foot two-story ICF house (whose hypothetical sales price amounts to $180,000) comes at a $7,000 extra cost. Such additional cost has a negligible impact on the overall cost if compared to the plenty of guarantees and long-term benefits this technology brings. It should also be noted that this premium is directly offset by the money saved in energy/utility bills, reduced insurance premiums, and so on.

As far as further advantages are concerned, the study offers plenty of experiences and opinions given by former or current ICF building users through survey reports. It is worth noting that a few occupants, who have experienced tornado survival in Iowa and Illinois, stated that no damage occurred in their houses unlike in most traditional buildings.

Lastly, the study provides further evaluations on ICFs benefits, such as structural integrity following earthquakes, fire resistance, and durability. This leads us to the conclusion that the ICFs "package deal" is more cost-effective than traditional framing.

SUMMARY SHEET:

Title:

Costs and benefits of Insulating Concrete Forms for residential construction

Authors:

NAHB Research Center, Inc.Upper Marlboro, MD

Prepared for:

Department of Civil Engineering, Portland State University

Publication type:

Research contribution

Date of publication:

2001

Publisher:

ISBN:

Content summary:

Cost evaluation of ICF systems

Text length and type:

30 folders containing pictures and graphs

URL for the full text:

https://www.huduser.gov/Publications/PDF/icfbenefit.pdf

3.6.3) Assessment on the usage of Insulated Concrete Form in United Arab Emirates construction industry

This article focuses on the increasing demand for sustainable and environmentally-friendly buildings in the Emirates, where new regulations on green building have been recently introduced. This will inevitably lead towards the use of greener construction methods such as ICFs.

The article starts by reporting the results of a survey on the Arabic operators' knowledge of ICFs and experience in the field. The results showed that this technology is scarcely known in the Emirates, probably due to the little publicity this innovation has received.

Afterwards, the operators were provided with a few American scientific papers on several topics concerning the advantages of the ICF constructive system.

Throughout the survey, many technical categories ranging from builders to engineers were asked about the ICFs potential in the Arabic marketplace. The results showed that according to the majority, ICFs may gain in popularity thanks to the freshly introduced regulations on sustainable building.

Lastly, the study express the Authors' intention to undertake and spread further in-depth analysis on the overall costs of ICF constructions, including the initial, usage, and maintenance costs. In order to encourage the spread of this system, it is indeed necessary to provide evidence on the lower costs of ICF systems over traditional building.

Conclusion

The building industry is undergoing a peculiar historical period characterized by a decade-long economic crisis and newly introduced regulations whose executive and design standards are increasingly high. It really is a time of change for this industry… In such a scenario, our well-informed buyers pay closer attention to once overlooked details. One of these are effective building safety measures.

The need for "effective" measures is mainly due to the fact that the many existing technical regulations do not always make a full guarantee for safety, as people have not failed to notice. Many are the recent dramatic reports of buildings whose structure ran into serious problems after a disaster.

We are duty-bound to consider as follows; why do we keep building according to traditional systems (wooden or reinforced concrete frames) whose effective safety can only be guaranteed by oversizing them?

The benefits offered by ICF systems come as an answer. Indeed, their monolithic over-strengthened reinforced concrete walls, interlocked and "double-walled", do guarantee the safety of its occupants. If properly designed, they also avoid damage even after massive earthquakes. Not only does this make the building occupants more "reassured", but it also reduces the social security contributions following a disaster.

Furthermore, this comes at lower or comparable cost (which mostly depends on the geographic areas of reference) than that of traditional or different "innovative" construction methods.

This exhausts the subject as far as the structural safety is concerned. However, there is still plenty of aspects left undealt with…

After the analysis of the fifty works on ICFs reviewed in this volume, the scientific evidence leaves no space to doubt; in areas that are rather seismic or under exceptional loadings, there is no better option than ICFs.

I would like to conclude by quoting a builder who has recently asked me: "If this were true, then why wouldn't ICF systems be imposed by law?" Well, since IT IS TRUE, I cannot answer this. Or rather, I'd prefer not to.

It is my hope that this collection of summary reports will provide a useful contribution to the spread of information on this topic. Also, I hope this publication will foster dialogue between the aforementioned researchers.

The credit for this work goes to them.

So does the credit for paving the way for the manufacturers that can now undergo further analysis on these present-day technological solutions or seek to implement them.

Youcanprint

Printed in March 2020